T0226250

Synthesis Lectures on Electrical Engineering

This series of short books covers a broad spectrum of titles of interest in electrical engineering that may not specifically fit within another series. Books will focus on fundamentals, methods, and advances of interest to electrical and electronic engineers.

Farzin Asadi

Electric Circuits Laboratory Manual

Farzin Asadi
Department of Electrical and Electronics
Engineering
Maltepe University
Istanbul, Turkey

ISSN 1559-811X   ISSN 1559-8128   (electronic)
Synthesis Lectures on Electrical Engineering
ISBN 978-3-031-24554-1   ISBN 978-3-031-24552-7   (eBook)
https://doi.org/10.1007/978-3-031-24552-7

This Springer imprint is published by the registered company Springer Nature Switzerland AG
The registered company address is: Gewerbestrasse 11, 6330 Cham, Switzerland

Preface

This is a book for a lab course meant to accompany, or follow, any first course in electric circuit analysis. It has been written for sophomore or junior electrical and computer engineering students who are taking their first lab, either concurrently with their first electric circuit analysis class or following that class. This book is appropriate for non-majors, such as students in other branches of engineering and in physics, for which electric circuits is a required course or elective and for whom a working knowledge of electric circuits is desirable.

This book has the following objectives:

1. To support, verify, and supplement the theory; to show the relations and differences between theory and practice.
2. To teach measurement techniques.
3. To convince students that what they are taught in their lecture classes is real and useful.
4. To help make students tinkerers and make them used to asking "what if" questions.

This book contains 33 experiments which help the reader to explore the concepts studied in the classroom. Here is a brief summary of the chapters and appendixes:

Chapter 1 introduces the commonly used measurement devices that are used during the experiments and breadboard to the reader.

Chapter 2 studies the resistors. This chapter contains 9 experiments.

Chapter 3 studies the details of measurement with Digital Multi Meters (DMM). This chapter contains 7 experiments.

Chapter 4 studies some of the important circuit theorems like Kirchhoff's Voltage Law (KVL), Kirchhoff's Current Law (KCL), nodal analysis, mesh analysis, and Thevenin equivalent circuit. This chapter contains 5 experiments.

Chapter 5 studies the first order (RC and RL) and second order (series and parallel RLC) circuits. This chapter contains 4 experiments.

Chapter 6 studies the DC and AC steady state behavior of electric circuits. Frequency response of filters are studied in this chapter as well. This chapter contains 5 experiments.

Chapter 7 studies magnetic coupling and transformers. This chapter contains 3 experiments.

Appendix A shows how to draw different types of graphs with MATLAB®.

Appendix B reviews the concept of Root Mean Square (RMS).

I hope that this book will be useful to the readers, and I welcome comments on the book.

Istanbul, Turkey   Farzin Asadi
                                                                      farzinasadi@maltepe.edu.tr

Contents

Commonly Used Labaratory Equipmentes

1

1.1 Introduction

This chapter studies the most commonly used devices in the laboratory. Measurement devices that are used in different laboratories are made by different companies. Working with each device has its own details. Therefore, you are encouraged to read the user manual of devices that you will use during the experiments. Another good reference is your laboratory instructor. Studying the reference [1] is highly recommended as well.

1.2 Digital Multi Meter (DMM)

Digital multimeters are measuring instruments that can measure quantities such as voltage, current, and resistance. Measured values are shown on a digital display, allowing them to be read easily and directly, even by first-time users.

Study the user manual of the DMM that you will use in the experiments. Ensure that you are able to do the followings:

(a) Measurement of resistance.
(b) Measurement of AC/DC voltages.
(c) Measurement of AC/DC currents in the range of Amps.
(d) Measurement of AC/DC currents in the range of milli/micro Amps.

Ask your laboratory instructor to make an explanation to you if you are not able to do one or more of the above tasks.

© The Author(s), under exclusive license to Springer Nature Switzerland AG 2023
F. Asadi, *Electric Circuits Laboratory Manual*, Synthesis Lectures
on Electrical Engineering, https://doi.org/10.1007/978-3-031-24552-7_1

1.3   Function Generator (Signal Generator)

A function generator (sometime is called signal generator) is used to generate different types of electrical waveforms over a wide range of frequencies. Some of the most common waveforms produced by the function generator are the sine wave, square wave, triangular wave and saw tooth shapes.

The FG's are divided into two groups: Analog FG's and Direct Digital Synthesis (DDS) FG's. As the name suggests, the analog FG's, uses the analog circuits in order to produce the output waveform. DDS FG's use digital circuits (i.e. a microprocessor) in order to produce the output waveforms. Accuracy of DDS signal generators is better than analog signal generators.

Beside the standard waveforms (i.e. sinusoidal, square, triangular and saw tooth), some DDS FG's are able to produce arbitrary waveforms. These type of FG's are called Arbitrary Waveform Generator (AWG). They have software which permits you to draw the waveform that you want. After drawing the waveform in the software environment, the hardware of AWG produces the waveform for you.

Output of function generator is connected to the circuit under test with the aid of a cable (Fig. 1.1). Black wire is connected to the ground of the signal generator as shown in Fig. 1.2. R1 shows the output resistance of the signal generator which is generally 50 Ω.

Fig. 1.1 Signal generator cable

Fig. 1.2 Simple model for
signal generator

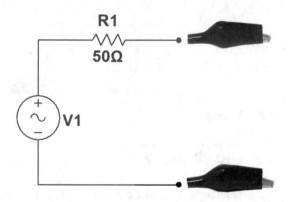

Study the user manual of the function generator that you will use in the experiments. Ensure that you are able to do the followings:

(a) Generation of a sinusoidal signal with amplitude of 5 V and frequency of 50 Hz, i.e., $v(t) = 5 \times \sin(2\pi \times 50 \times t)V$.
(b) Generation of a sinusoidal signal with amplitude of 5 V, frequency of 50 Hz and average value of 3 V, i.e., $v(t) = 3 + 5 \times \sin(2\pi \times 50 \times t)V$.
(c) Generation of a sinusoidal signal with peak value of 50 mV, i.e., $v(t) = 0.05 \times \sin(2\pi \times 50 \times t)V$.
(d) Generation of a triangular wave with frequency of 50 Hz and peak value of 5 V (Fig. 1.3).

Fig. 1.3 Sample triangular
wave

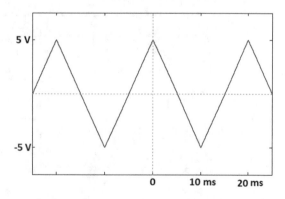

(e) Generation of a pulse with frequency of 50 Hz and duty cycle of 25% (Fig. 1.4).

Ask your laboratory instructor to make an explanation to you if you are not able to do one or more of the above tasks.

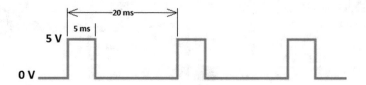

Fig. 1.4 Sample pulse

1.4   Oscilloscope

An oscilloscope is an instrument that graphically displays electrical signals and shows how those signals change over time. Scientists, engineers, physicists, repair technicians and educators use oscilloscopes to see signals change over time. These days generally digital scopes are used in laboratory.

Study the user manual of the oscilloscope that you will use in the experiments. Ensure that you are able to do the followings:

(a) Measurement of peak value of a signal.
(b) Measurement of period and frequency of a periodic signal.
(c) Use the cursors to read voltage or time difference between two points.
(d) Observing two waveforms simultaneously.
(e) Measurement of phase difference between two waveforms. Remember that you can measure the phase difference between two waveform easily with the aid of $\Delta\varphi = \frac{\Delta t}{T} \times 360°$ formula (Fig. 1.5).

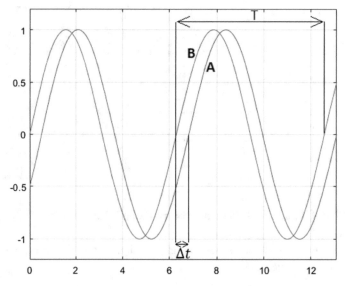

Fig. 1.5 Measurement of phase difference. B leads A by $\Delta\varphi = \frac{\Delta t}{T} \times 360°$

(f) Explain the difference between AC and DC coupling.
(g) Explain the functionality of X1/X10 switch on the probe.

Ask your laboratory instructor to make an explanation to you if you are not able to do one or more of the above tasks.

1.5   Power Supply

All the circuits require an energy source in order to work. The power supply (PS) is responsible for providing the required energy for the circuit. The power supply takes the AC electric energy from the grid and converts it into a DC voltage. Generally, they provide the voltages in the 0–30 V range. Generally, the output current could be up to 3 A. The outputs of a PS are called a "Channel". So, when we speak about a 3 channel PS, we mean a PS with three outputs. Generally, the outputs are variable and the user could set them to the desired value he/she wants. Generally, PS's have one regulated output with voltage of 5 V. This output is used to supply digital circuits. Remember that traditional digital circuits work with 5 V (However this voltage decreased to 3.3 V and even 1.1 V these days!). So, it is a good idea to use this fixed 5 V when you work with traditional digital circuits. You can connect a digital circuit to variable outputs of a PS. However, if you increase the voltage of that variable channel by mistake, then your circuit may be damaged. So, always use this fixed 5 V when you are working with traditional digital circuits.

Study the user manual of the power supply that you will use in the experiments. Ensure that you are able to do the followings:

(a) Generation of 12 V with maximum output current of 0.5 A.
(b) Generation of 12 V with maximum output current of 5 A (Use parallel mode).
(c) Generation of 40 V with maximum output current of 1 A (Use series mode).
(d) Generation of a symmetric voltage for instance +12 V and −12 V with maximum output current of 0.5 A.

Ask your laboratory instructor to make an explanation to you if you are not able to do one or more of the above tasks.

1.6   Breadboard

A breadboard is used to build and test circuits quickly. The breadboard has many holes into which circuit components like ICs and resistors can be inserted. A typical breadboard is shown in Fig. 1.6.

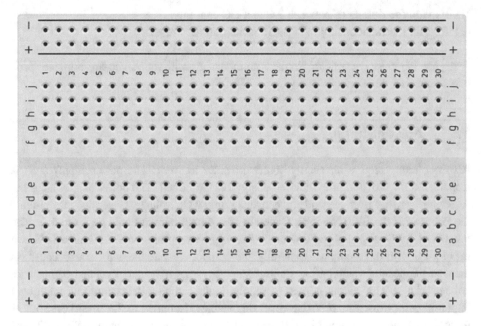

Fig. 1.6 Bread board

The breadboard has strips of metal which run underneath the board and connect the holes on the top of the board. The metal strips are laid out as shown below. Note that the top and bottom rows of holes are connected horizontally while the remaining holes are connected vertically (Figs. 1.7 and 1.8).

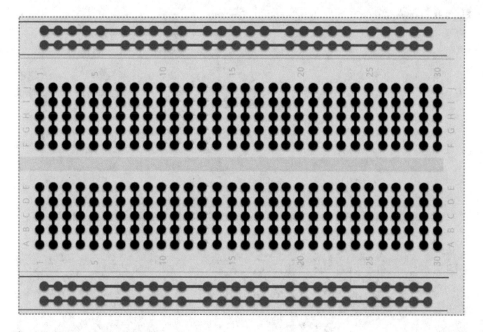

Fig. 1.7 Connections of breadboard

Fig. 1.8 Inside of a
breadboard

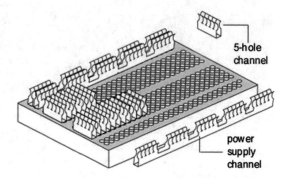

Let's study an example. Assume that want to make the circuit shown in Fig. 1.9 on breadboard.

Fig. 1.9 Sample circuit

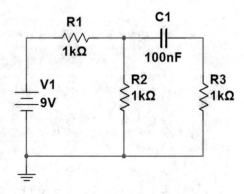

Let's start from the left side. Put the resistor R1 on the breadboard (Fig. 1.10).

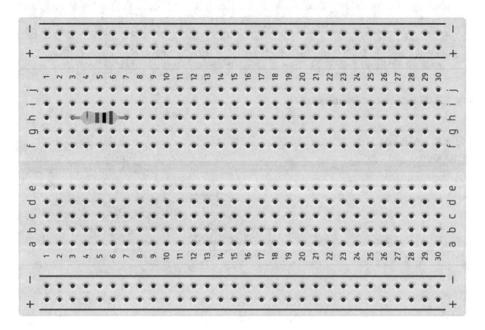

Fig. 1.10 Resistor R1 is added

Put the resistor R2 (Fig. 1.11).

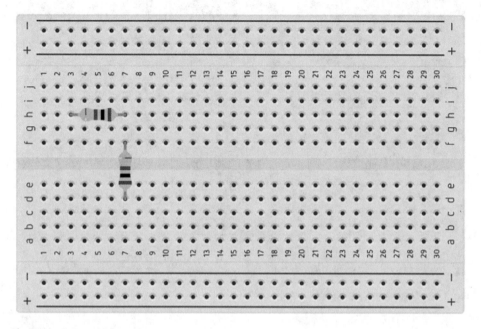

Fig. 1.11 Resistor R2 is added

Put the capacitor C1 (Fig. 1.12).

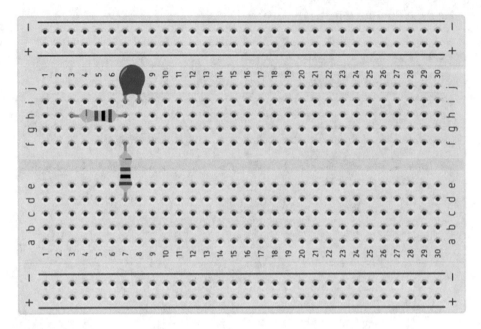

Fig. 1.12 Capacitor C1 is added

Put the resistor R3 (Fig. 1.13).

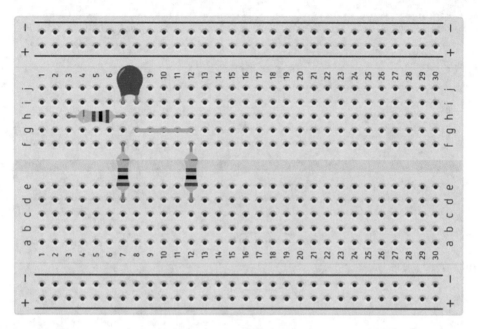

Fig. 1.13 Resistor R3 is added

Connect the resistor R1 to the positive voltage rail and R2 and R3 to the negative voltage rail (Fig. 1.14).

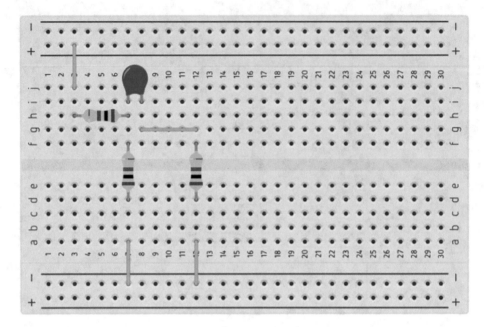

Fig. 1.14 Power supply connection are made

Connect the power supply to the breadboard (Fig. 1.15).

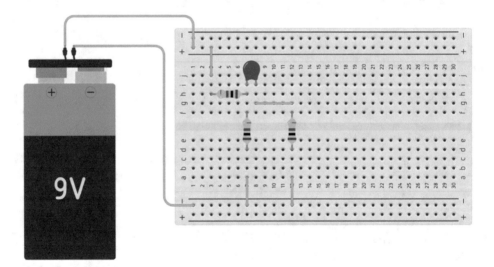

Fig. 1.15 Voltage source is added to the circuit

Measurement with cell phone

You can convert your cell phone, tablet or even your smart watch into a digital multimeter, digital storage oscilloscope (DSO) or a logger with the aid of Pokit meter (Figs. 1.16 and 1.17) or Pokit pro (Fig. 1.18). More information can be found on https://www.pokitinno vations.com/.

Fig. 1.16 Pokit meter

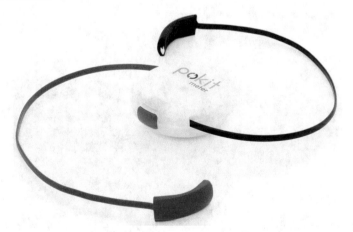

Fig. 1.17 Pokit meter

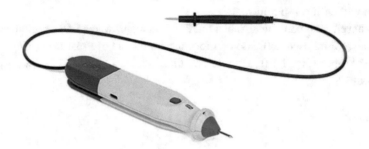

Fig. 1.18 Pokit pro

Reference for Further Study

1. Asadi F., Eguchi K., Electronic Measurement: A Practical Approach, Springer, 2021

Resistors

2

2.1 Introduction

Resistors play an important role in electric and electronic circuits. In this chapter you will learn about different types of resistors, Ohm' law, how to measure the resistance using Digital Multi Meter (DMM) and how to convert a current signal into a voltage signal. This chapter includes 9 experiments.

2.2 Tolerance of Resistors

2.2.1 Introduction

Any component used in a circuit has a tolerance. The tolerance of a component is a measure of accuracy and indicates how much the measured actual value is different from its nominal expected value. For instance, a resistor with nominal value of 1 kΩ and tolerance of 5% may have any value between 1 k$\Omega \times 0.95 = 0.95$ kΩ and 1 k$\Omega \times 1.05 = 1.05$ kΩ.

In this experiment we will study the tolerance of resistors. Let's see how value of a resistor is read. In big bulky resistors there is enough space to write the nominal value and tolerance code using numbers and alphabet letters. For instance, the resistor shown in Fig. 2.1 has nominal value of 0.47 Ω and is capable to dissipate 5 W of heat. Therefore, maximum value of 3.26 A can pass through the resistor. According to Table 2.1, letter J shows that this resistor has tolerance of 5%. Therefore, actual value of the resistor may be between 0.95×0.47 $\Omega = 0.45\Omega$ and 1.05×0.47 $\Omega = 0.49$ Ω.

© The Author(s), under exclusive license to Springer Nature Switzerland AG 2023
F. Asadi, *Electric Circuits Laboratory Manual*, Synthesis Lectures
on Electrical Engineering, https://doi.org/10.1007/978-3-031-24552-7_2

Fig. 2.1 High wattage resistor

Table 2.1 Tolerance
associated with letters

Letter	Tolerance associated with the letter
B	0.1%
C	0.25%
D	0.5%
F	1%
G	2%
J	5%
K	10%
M	20%

Using color bands is another commonly used method to show the value of resistors (Fig. 2.2). This method is used for small non-bulky resistors. Number of color bands may be 4 or 5. In resistors with 4 color bands, 3 bands show the value of resistor and 1 band shows the tolerance. In resistors with 5 color bands, 4 bands show the value of resistor and 1 band shows the tolerance. Values associated with each color code is shown in Table 2.2. It is a good idea to memorize this table.

Fig. 2.2 Value of resistor is
shown with color bands

Table 2.2 Digits associated with colors

Color	Digit associated with the color
Black	0
Brown	1
Red	2
Orange	3
Yellow	4
Green	5
Blue	6
Violet	7
Grey	8
White	9

In resistors with 4 color bands, first and second color bands show the first and second digits and third color band shows the multiplier. Fourth color band shows the tolerance and usually takes the colors shown in Table 2.3. For instance, assume that third band is green. In this case first two digits must be multiplied with $10^5 = 100000$. Note that when third color is gold, the first two digits must be divided by 10.

Table 2.3 Tolerance associated with colors

Color	Tolerance associated with the color
Golden	5%
Silver	10%

For instance, a resistor with brown–green–red–gold color bands shows a $15 \times 100 = 1500 \ \Omega = 1.5$ kΩ resistor with tolerance of 5%. Actual value of such a resistor changes from 1.5 kΩ $\times 0.95 = 1.43$ kΩ to 1.5 kΩ $\times 1.05 = 1.58$ kΩ. Table 2.4 shows color bands of different resistors. In this table, for instance, 5R6 shows 5.6 Ω, 8K2 shows 8.2 kΩ, 330 K shows 330 kΩ, 12 M shows 12 MΩ and 3M9 shows 3.9 MΩ.

Table 2.4 Color bands for different resistors

Band 1	Band 2	Band 3							
		Gold	Black	Brown	Red	Orange	Yellow	Green	Blue
Brown	Black	1R0	10R	100R	1K0	10K	100K	1MO	10M
Brown	Red	1R2	12R	120R	1K2	12K	120K	1M2	12M
Brown	Green	1R5	15R	150R	1K5	15K	150K	1M5	15M
Brown	Grey	1R8	18R	180R	1K8	18K	180K	1M8	18M
Red	Red	2R2	22R	220R	2K2	22K	220K	2M2	22M
Red	Violet	2R7	27R	270R	2K7	27K	270K	2M7	27M
Orange	Orange	3R3	33R	330R	3K3	33K	330K	3M3	33M
Orange	White	3R9	39R	390R	3K9	39K	390K	3M9	39M
Yellow	Violet	4R7	47R	470R	4K7	47K	470K	4M7	47M
Green	Blue	5R6	56R	560R	5K6	56K	560K	5M6	56M
Blue	Grey	6R8	68R	680R	6K8	68K	680K	6M8	68M
Grey	Red	8R2	82R	820R	8K2	82K	820K	8M2	82M

In resistors with 5 color bands, first, second and third color bands show the first, second and thirds digits and fourth color band shows the multiplier. Fifth color band shows the tolerance and usually takes the colors shown in Table 2.5. Note that when fourth color is gold, the first three digits must be divided by 10. For instance, green–brown–black–red–brown (Fig. 2.3) shows 51 kΩ with tolerance of 1%.

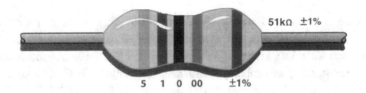

Fig. 2.3 Resistor with 5 color bands

Table 2.5 Tolerance associated with colors

Color	Tolerance associated with the color (%)
Brown	1
Red	2
Golden	5
Silver	10

A logical question is "How do I know which end of the resistor to start reading from?". Following clues help you to find answer for this question:

(A) Many resistors have some of the color bands grouped closer together or grouped toward one end. Hold the resistor with these grouped bands to your left. Always read resistors from left to right.
(B) Resistors never start with a metallic band on the left. If you have a resistor with a gold or silver band on one end, you have a 5% or 10% tolerance resistor. Position the resistor with this band on the right side and again read your resistor from left to right.
(C) Basic resistor values range from 0.1–10 MΩ. With that knowledge, realize that on a four-band resistor the third color will always be blue or less and on a five-band resistor, the fourth color will always be green or less.

The standard resistor values are organized into a set of series of values known as the E-series. E24 series is the most commonly used series. In E24 series, all of the values shown in Table 2.6 multiplied by 10^n where $-1 \leq n \leq 6$ are available. For instance, 0.27 Ω, 2.7 Ω, 27 Ω, 270 Ω, 2.7 kΩ, 27 kΩ, 270 kΩ and 2.7 MΩ are available in this series. E24 series has tolerance of 5%.

Table 2.6 E24 standard resistor series

1.0	1.1	1.2
1.3	1.5	1.6
1.8	2.0	2.2
2.4	2.7	3.0
3.3	3.6	3.9
4.3	4.7	5.1
5.6	6.2	6.8
7.5	8.2	9.1

In this experiment we will learn how to read the nominal value of a resistor using color codes and how to measure a resistor value with a DMM. We will study the effect of temperature increase on carbon resistors as well.

2.2.2 Procedure

Prepare 100 Ω, 1 kΩ, 10 kΩ and 100 kΩ resistors. Write the color bands in Table 2.7.

Table 2.7 Color bands of resistors under test

Nominal value	Color of first band	Color of second band	Color of third band	Color of fourth band (tolerance band)
100 Ω				
1 kΩ				
10 kΩ				
100 kΩ				

Fill the Table 2.8 based on the tolerance band of the resistors.

Table 2.8 Maximum and minimum values of resistors under test

Nominal value	Minimum value	Maximum value
100 Ω		
1 kΩ		
10 kΩ		
100 kΩ		

Use a DMM to measure the values of resistors and write them in the Table 2.9.

Table 2.9 Measured values for resistors under test

Nominal value	Value measured with DMM
100 Ω	
1 kΩ	
10 kΩ	
100 kΩ	

Ensure that measured values lies between the minimum and maximum values of Table 2.8. Now keep a hot soldering iron behind the resistors for 30 s and measure the resistances again. Write the measured values in Table 2.10. Compare the Table 2.10 with values measured in room temperature. What is the effect of temperature increase on the resistance?

Table 2.10 Measured values for heated resistors

Nominal value	Value measured with DMM
100 Ω	
1 kΩ	
10 kΩ	
100 kΩ	

2.3   Measurement of Low Resistances

2.3.1   Introduction

DMM probes have a resistance in the range of few tenth of the Ohm. When you want to measure low resistors (in the range of few Ohms) with DMM, the resistance of DMM probes is added in series to the resistor under test. So, the measurement is not so accurate.

In order to have an accurate measurement, you need to subtract the probe resistance from the value shown by the DMM. When you want to measure the big resistors (for instance in the range kΩ), such a subtraction is not necessary because value of the resistor under test is quite bigger than the resistance of the probes. In this experiment measurement of low resistors is studied.

2.3.2   Procedure

Prepare a low value resistor. For instance, a 2.2 or 3.3 Ω resistor is good. Use the tolerance band to calculate the minimum and maximum values of the resistor under test and call them Rmin and Rmax, respectively.

Now connect the DMM leads to the resistor under test and measure its value (Fig. 2.4). Call the measured value R1.

Fig. 2.4 Resistor is connected to the DMM

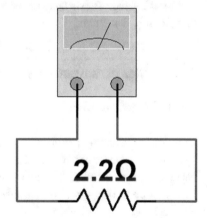

Connect the test leads together in order to measure the resistance of leads (Fig. 2.5). Write the value shown on the display and call it R2.

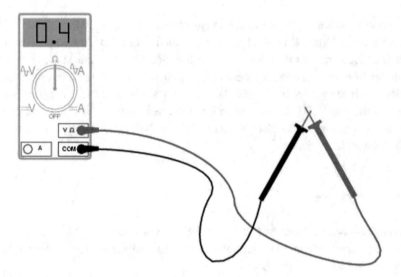

Fig. 2.5 Probes are shorted together. In this figure resistance of leads is 0.4 Ω

Correct value of resistance under test is R1–R2. Calculate the R = R1–R2 and ensure that Rmin < R < Rmax.

2.4   Measurement of Very Low Resistances

2.4.1   Introduction

Very low resistances (for instance, resistance of cables) can't be measured with DMM. Very low resistances can be measured with the aid of Ohm's law: We need to pass a known current through the resistor under test and measure the voltage drop across it. Ratio of voltage drop to the current through the resistor gives the resistance. Measurement of very low resistances is studied in this experiment.

2.4.2   Procedure

Prepare the circuit shown in Fig. 2.6. Value of R1 and V1 are selected such that a few Amps pass through the wire under test. For instance, for V1 = 12 V and R = 6.8 Ω, a current around 12/6.8 = 1.76 A pass through the wire (R1 power rating must be high enough to withstand such a current). After passing the current through the wire, a voltage in the range of few milli Volts is shown by the milli voltmeter. According to Ohm's law, the resistance of the wire under test is equal the value shown by milli Voltmeter divided by the value shown by the Ammeter.

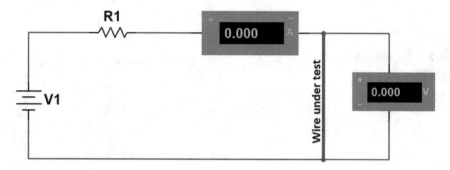

Fig. 2.6   Measurement of a very low resistance

If you have access to a rheostat, you can use the circuit shown in Fig. 2.7 as well. In this figure R1 is a rheostat and permits us to change the circuit current. Don't decrease the value of R1 too much otherwise you make a short circuit the V1 terminals.

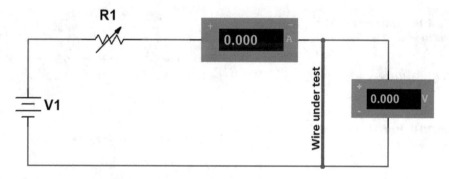

Fig. 2.7   Circuit current can be controlled with a rheostat

2.5   Series and Parallel Connection of Resistors

2.5.1   Introduction

You can use series and parallel connection of resistors in order to make a resistor which is not available in the table of standard resistors. Series and parallel connection of two resistors are shown in Figs. 2.8 and 2.9, respectively. Equivalent resistance for Figs. 2.8 and 2.9 is $R1 + R2$ and $\frac{R1 \times R2}{R1+R2}$, respectively.

Fig. 2.8 Series connected
resistors

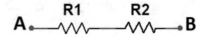

Fig. 2.9 Parallel connected
resistors

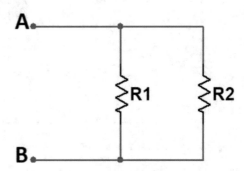

Series and parallel connection of capacitors are shown in Figs. 2.10 and 2.11, respectively. Equivalent capacitance for Figs. 2.10 and 2.11 is $C1 + C2$ and $\frac{C1 \times C2}{C1+C2}$, respectively.

Fig. 2.10 Series connected
capacitors

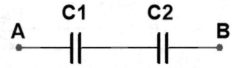

Fig. 2.11 Parallel connected
capacitors

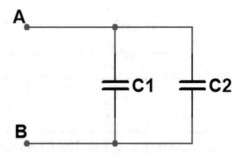

Series and parallel connection of inductors are shown in Figs. 2.12 and 2.13, respectively. Equivalent inductance for Figs. 2.12 and 2.13 is $\frac{L1 \times L2}{L1+L2}$ and $L1 + L2$, respectively. Note that there is no magnetic coupling between L1 and L2 in Figs. 2.12 and 2.13.

In this experiment series and parallel connection of components are studied.

Fig. 2.12 Series connected
inductors

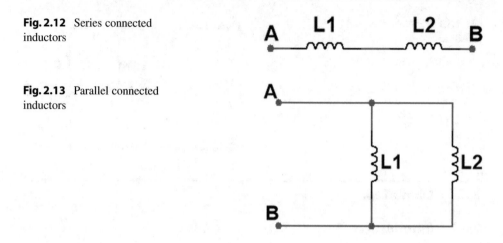

Fig. 2.13 Parallel connected
inductors

2.5.2 Procedure

Prepare a 1 kΩ and 10 kΩ resistor. Measure the resistance of resistors with a DMM and write the measured values. Connect the resistors in series (Fig. 2.14) and measure the resistance between A and B. Compare the measured value with R1 + R2. R1 and R2 show the measured values for 1 kΩ and 10 kΩ resistors, respectively.

Fig. 2.14 Series connected
resistors

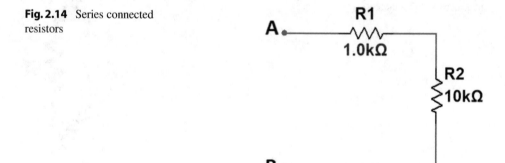

Now connect the resistors in parallel (Fig. 2.15) and measure the resistance between A and B. Compare the measured value with $\frac{R1 \times R2}{R1+R2}$.

If you have access to RLC meter you can verify the formulas for series/parallel inductor/capacitors as well.

Fig. 2.15 Parallel connected
resistors

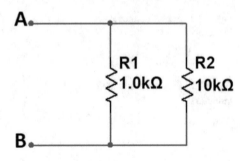

2.6   Ohm's Law

2.6.1   Introduction

Ohm's law states that the voltage across a conductor is directly proportional to the current flowing through it, provided all physical conditions and temperatures remain constant. Mathematically, this current–voltage relationship is written as: $V = R \times I$ (Fig. 2.16). In this equation, the constant of proportionality, R, is called Resistance and has units of Ohms, with the symbol Ω.

Fig. 2.16   V, I, and R, the
parameters of Ohm's law

Ohm's law is an empirical relation which accurately describes the conductivity of the vast majority of electrically conductive materials over many orders of magnitude of current. However, some materials do not obey Ohm's law; these are called non-ohmic.

2.6.2   Procedure

Prepare a 1 kΩ resistor and measure its value. Then make the circuit shown in Fig. 2.17.

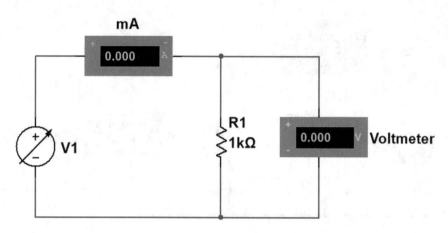

Fig. 2.17 Measurement of current and voltage of a resistor

Change the value of variable voltage source V1 in Fig. 2.17 in order to produce the currents shown in Table 2.11. Then write the corresponding voltage readings for each current and calculate the resistance using the Ohm's law for each row. The calculated resistance is almost constant.

Table 2.11 I–V data for resistor under test

Value read by milli Ampere meter (I)	Value read by Voltmeter (V)	$R = \frac{V}{I}$
1 mA		
3 mA		
5 mA		
7 mA		
9 mA		
11 mA		
13 mA		
15 mA		

When current through the resistor is 15 mA touch the resistor body with your fingers. The heat that you sense is the $p = RI^2 \approx 1k \times (15m)^2 = 225\,mW$ power dissipated in the resistor.

Now replace the resistor R1 in Fig. 2.17 with a small 12 V car lamp (Fig. 2.18). An Ammeter is used to measure the circuit current (Note that milli Ammeter is not used). Fill the Table 2.12. Is the lamp obeying the Ohm's law? Is it an Ohmic conductor?

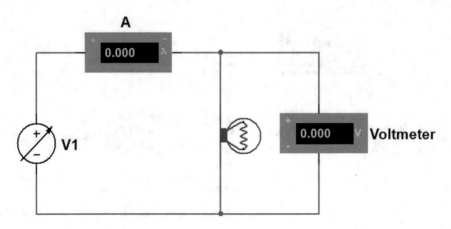

Fig. 2.18 Measurement of current and voltage of a small car lamp

	Value read by Ampere meter (I)	Value read by Voltmeter (V)	$R = \frac{V}{I}$
Table 2.12 I–V data for lamp under test			
		1	
		3	
		5	
		7	
		9	
		11	
		13	

2.7   Potentiometer

2.7.1   Introduction

A potentiometer is a manually adjustable variable resistor with 3 terminals (Fig. 2.19). Two of the terminals are connected to the opposite ends of a resistive element, and the third terminal connects to a sliding contact, called a wiper, moving over the resistive element. The potentiometer essentially functions as a variable resistance divider. The resistive element can be seen as two resistors in series (the total potentiometer resistance), where the wiper position determines the resistance ratio of the first resistor to the second resistor. If a reference voltage is applied across the end terminals, the position of the wiper determines the output voltage of the potentiometer (Fig. 2.20).

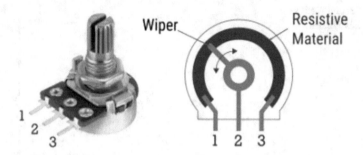

Fig. 2.19 A real potentiometer and its structure

Fig. 2.20 Voltage divider made with potentiometer

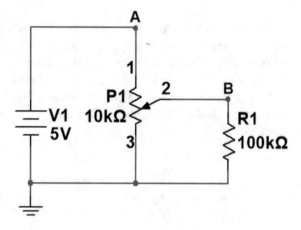

You can make a variable resistor with the aid of connections shown in Figs. 2.21 and 2.22.

Fig. 2.21 First method to make a variable resistor

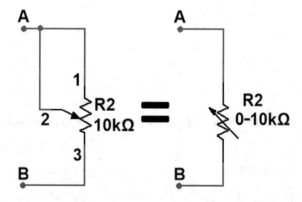

Fig. 2.22 Second method to
make a variable resistor

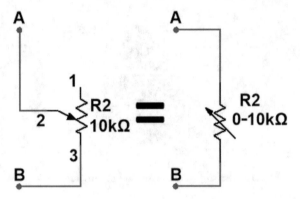

Method shown in Fig. 2.21 is better than the method shown in Fig. 2.22. In Fig. 2.22 we leave a pin open and that open pin can act as an antenna to absorb noise. So, if the potentiometer plays the role of a variable resistor in a sensitive electronic circuit, it is better to connect the unused pin to one of the side pins like Fig. 2.21.

This experiment studies the circuit shown in Fig. 2.23. Variable resistor R2 is made with a potentiometer. When value of R2 changes, the potential of point B changes from 0 V up to 2.5 V.

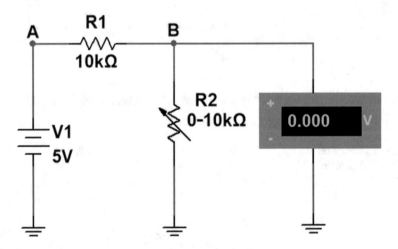

Fig. 2.23 Potential of point B changes with respect to value of R2

2.7.2 Procedure

Prepare a 10 kΩ potentiometer (Fig. 2.24). Connect the terminal 1 and 3 (Fig. 2.24) to a DMM (Fig. 2.25) and write the shown resistance. Compare this measured value with the nominal value printed on the potentiometer. Rotate the potentiometer shaft and observe that resistance shown by the DMM doesn't change.

Fig. 2.24 Labels 1–3 are given to potentiometer terminals

Fig. 2.25 Measurement of nominal value of the potentiometer

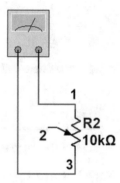

Connect the DMM to pin 2 and 3 (Fig. 2.26). Then rotate the potentiometer shaft. Write the maximum and minimum resistance values that DMM shows.

Fig. 2.26 Measurement of resistance between center terminal (wiper) and side terminal

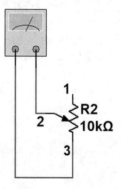

Now prepare the circuit shown in Fig. 2.27. Rotate the potentiometer and pay attention to voltage that voltmeter shows. Write the maximum and minimum voltage that voltmeter shows. The maximum voltage is obtained when the resistance between terminal 2 and 3 is maximum. The minimum voltage is obtained when the resistance between terminal 2 and 3 is minimum.

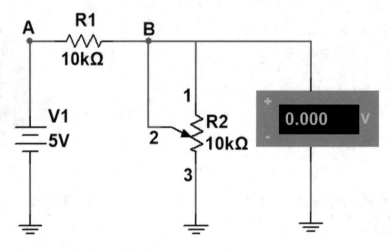

Fig. 2.27 Measurement of potential of point B

2.8   Light Dependent Resistor (LDR)

2.8.1   Introduction

Light Dependent Resistors (LDR) or photo resistors are electronic components that are often used in electronic circuit designs where it is necessary to detect the presence or the level of light. Resistance of LDR is a function of the light reaches to it.

A sample LDR is shown in Fig. 2.28. Symbol of LDR is shown in Fig. 2.29.

Fig. 2.28 Real LDR

Fig. 2.29 LDR symbols

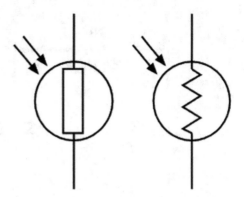

This experiment studies the behavior of LDR.

2.8.2 Procedure

Connect the LDR to the multimeter in order to measure its resistance (Fig. 2.30). Cover the LDR window with your hands. This decreases the amount of light which reaches to the LDR. Write the value which is shown by the multimeter. Now use a flashlight to shed light onto the LDR window. Write the value which is shown by the multimeter and compare it with the dark case.

Fig. 2.30 Measurement of LDR resistance

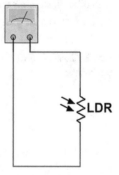

In Fig. 2.30 we used a digital multimeter to measure the resistance of the LDR. Now we will use an indirect method to measure the resistance of the LDR. Make the circuit shown in Fig. 2.31. Potential of point B changes with respect to the amount of light reaches to the LDR. Cover the LDR window with your hands and measure the voltage of point B. Use the $\frac{R_{LDR}}{R_1+R_{LDR}} \times V_1 = V_B \Rightarrow R_{LDR} = \frac{V_B}{V_1-V_B} R_1$ formula to calculate the resistance of the LDR. Then shed light onto the LDR window using a flashlight and measure the voltage of point B again. Calculate the resistance of the LDR for this case and compare it with the previous dark case.

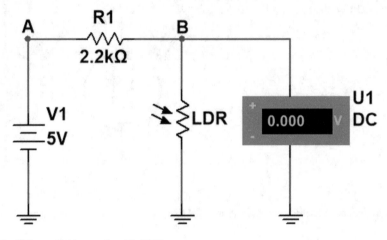

Fig. 2.31 Voltage divider made with LDR

2.9   Thermistor

2.9.1   Introduction

A thermistor is a resistor whose resistance is dependent on temperature (Fig. 2.32). The
term is a combination of "thermal" and "resistor". It is made of metallic oxides, pressed
into a bead, disk, or cylindrical shape and then encapsulated with an impermeable material
such as epoxy or glass.

Fig. 2.32 NTC type
thermistor

There are two types of thermistors: Negative Temperature Coefficient (NTC) and Positive Temperature Coefficient (PTC). With an NTC thermistor, when the temperature increases, resistance decreases. Conversely, when temperature decreases, resistance increases (Fig. 2.33). This type of thermistor is used the most.

A PTC thermistor works a little differently. When temperature increases, the resistance increases, and when temperature decreases, resistance decreases (Fig. 2.34). This type of thermistor is generally used as a fuse.

Fig. 2.33 Behavior of NTC type thermistors

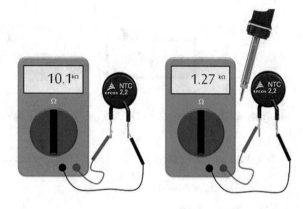

Fig. 2.34 Behavior of PTC type thermistors

International Electrotechnical Commission (IEC) and ANSI symbol for PTC and NTC are shown in Figs. 2.35 and 2.36, respectively.

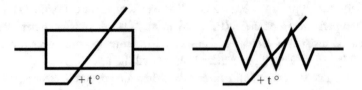

Fig. 2.35 Left: IEC symbol for PTC. Right: ANSI symbol for PTC

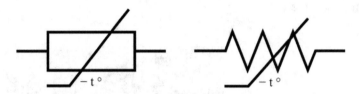

Fig. 2.36 Left: IEC symbol for NTC. Right: ANSI symbol for NTC

Behavior of thermistors are studied in this experiment.

2.9.2 Procedure

Connect the available thermistor to the DMM and measure its room temperature resistance. Now put a hot soldering iron behind the thermistor and measure its resistance. Based on the increase or decrease determine the type (PTC or NTC) of the thermistor.

Search the internet to find the datasheet of your thermistor and try to approximate the temperature of your soldering iron if the datasheet showed the resistance versus temperature graph.

2.10 Observing the Current Waveform

2.10.1 Introduction

Sometimes you need to observe a current signal. In order to be able to observe a current signal on the oscilloscope screen, you need to convert it into a voltage signal because oscilloscopes are able to show voltage signals only.

Let's study an example. Assume that we want to observe the current drawn by a computer fan from a 12 V power supply (Fig. 2.37).

Fig. 2.37 Current I is draw by
the fan

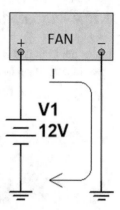

One way to convert the current signal into voltage signal is to put a small resistor in series with the load (Fig. 2.38). When the current pass through the sense resistor, it makes the voltage drop $V_{sense} = R_{sense} \times i(t)$ which is proportional with current pass through it. Wattage of the R_{sense} must be big enough to withstand the current pass through it. R_{sense} must be small enough to avoid a considerable voltage drop across it as well. This condition helps us to ensure that nearly all the input voltage reaches the load itself.

Fig. 2.38 Rsense act as a
current to voltage converter

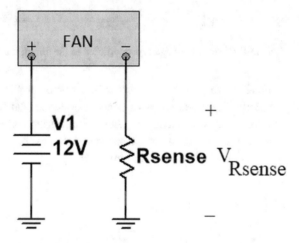

Conversion of current signal into voltage signal with the aid of sense resistor is studied in this experiment.

2.10.2 Procedure

Make the circuit shown in Fig. 2.39. Set the oscilloscope to see the voltage waveform across the Rsense. Use Ohm's law to measure the maximum current pass through the fan.

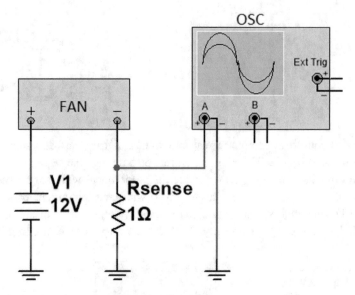

Fig. 2.39 Observed voltage is proportional to the circuit current

Note: If you don't have access to a computer fan, you can use the setup shown in Fig. 2.40. Signal generator V1 generates a sinusoidal signal with amplitude value of 5 V and frequency of 1 kHz. RO shows the output resistance of the signal generator V1. Voltage observed on the oscilloscope screen must be divided by 4.7 in order to obtain the equivalent current for each instant of time.

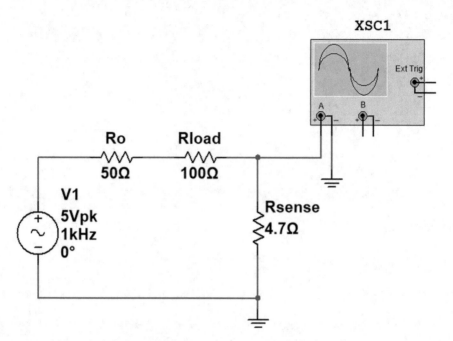

Fig. 2.40 A replacement for setup shown in Fig. 2.39

Reference for Further Study

1. Asadi F., Eguchi K., Electronic Measurement: A Practical Approach, Springer, 2021.

Digital Multi Meter

3

3.1 Introduction

This chapter focused on Digital Multi Meter (DMM) which is one of the important tools in the laboratory. This chapter shows how to measure the internal resistance of the DMM in the voltage/current measurement mode, how to measure the test current of a DMM in resistance measurement mode, how to determine whether the device is a True RMS device and how to obtain the frequency response of your DMM. This chapter contains 7 experiments.

3.2 Internal Resistance of DMM in Voltage Measurement Mode

3.2.1 Introduction

The internal resistance of the ideal voltmeter is infinity since it should not allow any current to flow through the voltmeter. Voltmeter measures the potential difference; it is connected in parallel.

A real voltmeter has a big (usually in the 10 MΩ range) but non-infinity internal resistance. Therefore, a real voltmeter loads the circuit under measurement. In this experiment internal resistance of a real voltmeter is measured.

3.2.2 Procedure

Prepare a 1 MΩ resistor and measure its resistance with DMM. Then make the circuit shown in Fig. 3.1.

© The Author(s), under exclusive license to Springer Nature Switzerland AG 2023   41
F. Asadi, *Electric Circuits Laboratory Manual*, Synthesis Lectures
on Electrical Engineering, https://doi.org/10.1007/978-3-031-24552-7_3

Fig. 3.1 1 MΩ resistor is
connected in series with
voltage source V1

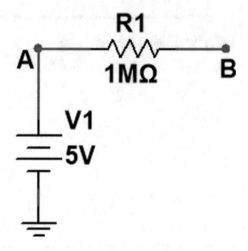

Measure the voltage of node A with a DMM (Fig. 3.2).

Fig. 3.2 Measurement of
potential of node A

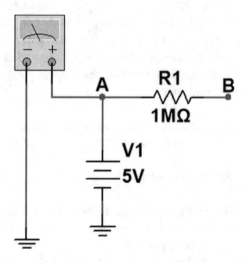

Measure the voltage of node B with a DMM (Fig. 3.3).

Fig. 3.3 Measurement of potential of node B

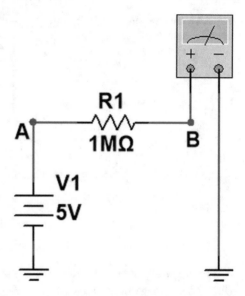

Use the $\frac{R_{VM}}{R_{VM}+R_1}V_A = V_B \Rightarrow R_{VM} = \frac{V_B}{V_A-V_B}R_1$ to calculate the internal resistance of the voltmeter. R_{VM}, R_1, V_A and V_B show the internal resistance of voltmeter, measured resistance for 1 MΩ resistor, voltage measured for node A and voltage measured for node B, respectively.

3.3   Internal Resistance of DMM in Current Measurement Mode

3.3.1   Introduction

The internal resistance of an ideal ammeter will be zero since it should allow current to pass through it. The ammeter is connected in series in a circuit to measure the current flow through the circuit.

A real ammeter has a small (usually in the 0.1 Ω range) but non-zero internal resistance. Therefore, a real ammeter loads the circuit under measurement. In this experiment internal resistance of a real ammeter is measured.

3.3.2   Procedure

Make the circuit shown in Fig. 3.4. The Ammeter in this figure is the Ammeter that we want to measure its internal resistance.

Fig. 3.4 Measurement of
circuit current

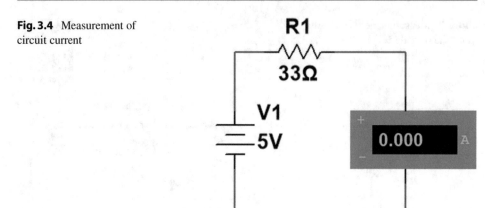

Prepare a separate voltmeter and measure the voltage drop across the Ammeter
(Fig. 3.5). Voltage drop across the Ammeter is in the milli Volt range. Internal resistance of the Ammeter (R_{AM}) can be calculated using the Ohm's law: $R_{AM} = \frac{V}{I}$. Note
that V and I show the voltmeter reading and Ammeter reading, respectively.

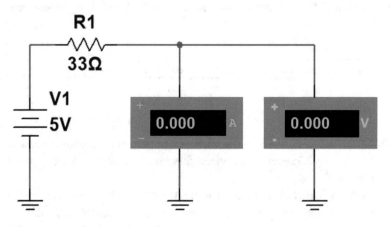

Fig. 3.5 Measurement of voltage drop of ammeter

3.4   Test Current of DMM in Resistance Measurement Mode

3.4.1   Introduction

A DMM uses a small test current to measure the resistance of the resistor under test (Fig. 3.6). The value of current is known. Therefore, the value of resistance equals to the voltage drop across the resistor under test divided by the current through it.

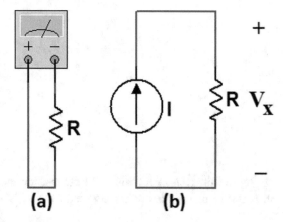

Fig. 3.6 a A resistor R is connected to a DMM in Ohm measurement mode **b** Equivalent circuit for (**a**)

In this experiment we want to measure the value of test current, i.e. the current that pass through the resistor under test during the resistance measurement. Note that DMM's use different test currents for different values of resistors. This is the job of DMM's CPU to select the best value for test current.

3.4.2   Procedure

Make the circuit shown in Fig. 3.7. The milli ammeter measures the current through the resistor under test. Write the current that is passed through the 1 kΩ resistor.

Fig. 3.7 Measurement of
current passed through the
resistor under test

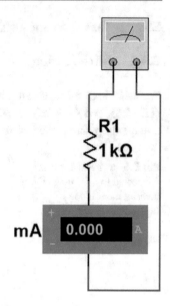

Replace the resistor R1 with a 22 kΩ resistor and pay attention to the current through
it. Is it the same as the test current used for 1 kΩ resistor?

3.5   DC Component (Average Value) Measurement with DMM

3.5.1   Introduction

If you apply a periodic signal to a DMM in DC voltage/current measurement mode, the
DC component (average value) will be shown. Remember that DC component (average
value) of a periodic signal $f(t+T) = f(t)$ is defined as $\frac{1}{T}\int_{t_0}^{t_0+T} f(\tau)d\tau$. t_0 is an arbitrary
value and T shows the period.

Let's study an example. If you apply the voltage shown in Fig. 3.8 to a DC voltmeter,
it will show 5 V. If you apply the voltage shown in Fig. 3.9 to a DC voltmeter, it will show
2.5 V since the average value of shown waveform is 2.5 V: $\frac{1}{10}\int_0^{10m} f(t)dt = \frac{5\times5m}{10m} =$
2.5 V.

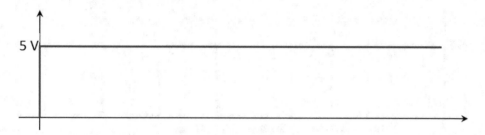

Fig. 3.8 A purely DC waveform

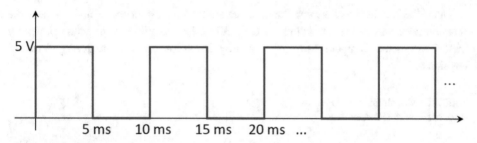

Fig. 3.9 Square waveform

3.5.2   Procedure

Connect the function generator to an oscilloscope (Fig. 3.10) and produce the waveform shown in Fig. 3.11. Average value of this waveform is $\frac{1}{T}\int_0^T f(t)dt = \frac{5\times 5m}{10m} = 2.5\,\text{V}$.

Fig. 3.10 Function generator
is connected to an oscilloscope

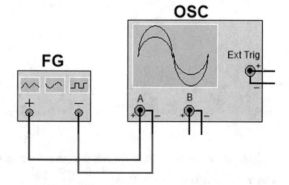

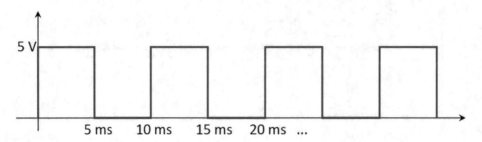

Fig. 3.11 Square waveform generated by function generator FG in Fig. 3.10

Open the connection between function generator and oscilloscope and connect the function generator to the DC voltmeter (Fig. 3.12). Pay attention to the value shown by the DC voltmeter. Compare the value shown by DC voltmeter with the average value of the signal.

Fig. 3.12 Measurement of DC component with DC voltmeter

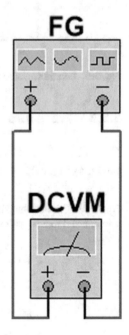

Repeat the above procedure for the waveform shown in Fig. 3.13. Average value of this waveform is $\frac{1}{T}\int_0^T f(t)dt = \frac{\frac{1}{2}\times 10\times 10m}{10m} = 5V$.

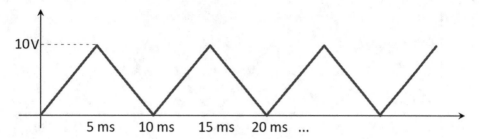

Fig. 3.13 Triangular waveform

3.6 RMS Measurement with DMM

3.6.1 Introduction

If you apply a periodic signal to a DMM in AC voltage/current measurement mode, the RMS of signal will be shown. Remember that RMS of a periodic signal $f(t+T) = f(t)$ is defined as $\sqrt{\frac{1}{T}\int_{t_0}^{t_0+T} f^2(\tau)d\tau}$. t_0 is an arbitrary value and T shows the period.

All of the DMM's are able to show the RMS value of purely sinusoidal signals (i.e., $V_m \sin(\omega t + \varphi)$) correctly. The RMS of non-sinusoidal signals (for instance the waveform shown in Figs. 3.14 and 3.15) is shown correctly only if the DMM is of "True RMS" type. In True RMS type DMM's, a CPU calculates the $\sqrt{\frac{1}{T}\int_{t_0}^{t_0+T} f^2(\tau)d\tau}$ formula. Price of True RMS type DMM's is higher than non-True RMS type DMM's.

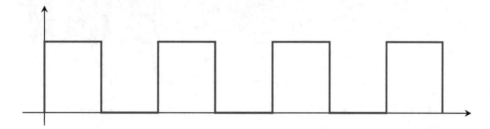

Fig. 3.14 Square waveform

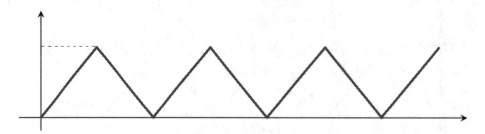

Fig. 3.15 Triangular waveform

AC voltage/current measurement is studied in this experiment.

3.6.2 Procedure

Connect the function generator to an oscilloscope (Fig. 3.16) and produce the waveform shown in Fig. 3.17. Note that RMS value of the waveform shown in Fig. 3.17 is
$\sqrt{\frac{1}{T}\int_0^T f(t)^2\,dt} = \sqrt{\frac{1}{20m}\int_0^{20m}(2\sin(100\pi t))^2\,dt} = \sqrt{\frac{1}{2\pi}\int_0^{2\pi}(2\sin(\alpha))^2\,d\alpha} = \frac{2}{\sqrt{2}} = 1.41\text{V}.$

Fig. 3.16 Function generator
FG is connected to an
oscilloscope

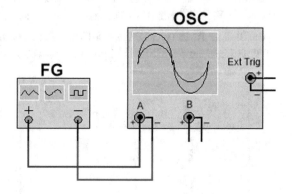

Fig. 3.17 Waveform
generated by function
generator FG in Fig. 3.16

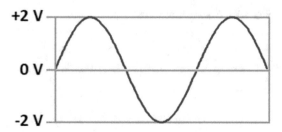

Open the connection between function generator and oscilloscope and connect the function generator to the AC voltmeter (Fig. 3.18). Pay attention to the value shown by the AC voltmeter. Compare the value shown by AC voltmeter with the RMS value of the signal. The equivalent circuit for Fig. 3.18 is shown in Fig. 3.19. Note that the 50 Ω resistor shows the output resistance of the function generator. Since the internal resistance of the AC voltmeter is much bigger than the output resistance of the function generator, nearly all of the input voltage drops across the AC voltmeter. Therefore, the AC voltmeter measures the RMS of V1.

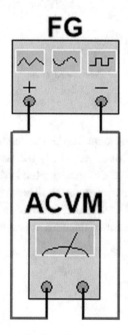

Fig. 3.18 AC voltmeter is connected to the function generator

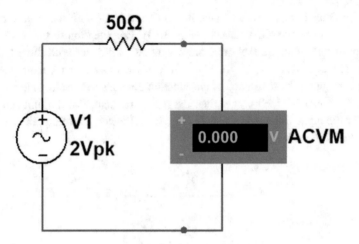

Fig. 3.19 Equivalent circuit for Fig. 3.18

Note: The RMS of a periodic voltage waveform can be calculated with the aid of a "Measure" menu of digital oscilloscopes. Study the oscilloscope manual or search the YouTube in order to learn how you can measure the RMS value with your oscilloscope.

Now replace the AC voltmeter with an AC milli ammeter (Fig. 3.20). Equivalent circuit of Fig. 3.20 is shown in Fig. 3.21. Note that internal resistance of the AC milli ammeter is around 0 Ω. Therefore, waveform of the current pass through the AC milli ammeter is similar to Fig. 3.22. RMS value of the current shown in Fig. 3.22 is $\frac{40\,\text{mA}}{\sqrt{2}} = 28.28\,\text{mA}$. Pay attention to the value shown by the AC milli ammeter. Compare the value shown by the AC milli ammeter with the RMS value of the current signal.

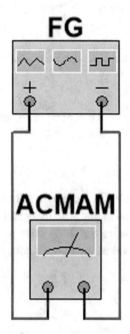

Fig. 3.20 AC milli ammeter is connected to the function generator

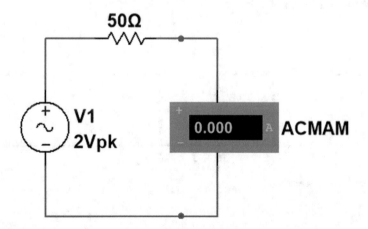

Fig. 3.21 Equivalent circuit for Fig. 3.20

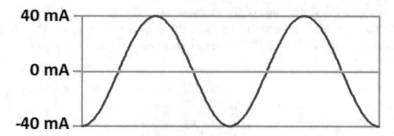

Fig. 3.22 Current waveform for Fig. 3.20

3.7   True RMS DMM's

3.7.1   Introduction

RMS value of non-sinusoidal signals is shown correctly only if the measurement device is of True RMS type. In this experiment you will apply non-sinusoidal signals to your DMM and see whether it is able to measure the RMS value of non-sinusoidal signals correctly. If your DMM shows the wrong value, then your device is not a True RMS type.

3.7.2   Procedure

Generate the waveforms shown in Figs. 3.23 and 3.24. RMS values of Figs. 3.23 and 3.24 are 3.53 and 5.77 V, respectively. Apply these waveforms to the AC voltmeter and see whether or not the device shows the expected correct values. If the results are not correct, the device is not a True RMS device.

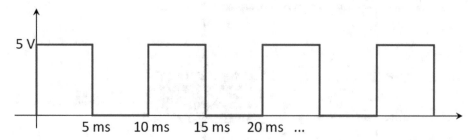

Fig. 3.23 Square waveform

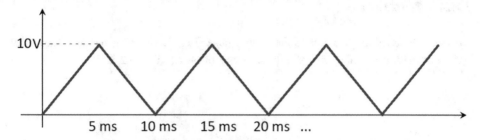

Fig. 3.24 Triangular waveform

3.8 Frequency Response of the AC Voltmeter and AC Ammeter

3.8.1 Introduction

You learned that not all the DMM's show the RMS value of non-sinusoidal signals correctly. In this experiment you will observe one of the other limitations of an AC voltmeter/ammeter. An AC measurement device (AC voltmeter or ammeter) is able to measure the RMS value of signals which their frequency falls in a specific range. RMS of signals which their frequency falls outside of this range will not be measured correctly.

The simple block diagram shown in Fig. 3.25 tries to show this concept. A real AC measurement device is modelled as series connection of a low-pass filter and an ideal AC measurement device. The signal pass through the low-pass filter first. Amplitude of the signal which reaches the ideal AC meter equals to the amplitude of input signal times the gain filter of filter at the frequency of the input. Gain of the low-pass filter shown in Fig. 3.25 is always less than or equals to 1. Therefore, the signal which reaches the ideal AC meter has a smaller amplitude and this cause the shown value to be less than the correct value. For instance, when input signal has frequency of $\omega_0 \frac{Rad}{s}$, gain of filter is $\left|\frac{1}{1+\frac{j\omega_0}{\omega_0}}\right| = \left|\frac{1}{1+j}\right| = 0.707$, therefore, the RMS value shown by the ideal AC meter is 0.707 times the correct RMS value.

Input signal ⟶ $H(s) = \dfrac{1}{1 + \dfrac{s}{\omega_0}}$ ⟶ Ideal AC meter ⟶

Fig. 3.25 Block diagram of a real voltmeter

In this experiment behavior of an AC measurement device to different frequencies are studied.

3.8.2   Procedure

Connect the function generator to the oscilloscope (Fig. 3.26) and generate a sinusoidal waveform with peak value of 2 V and frequency of 50 Hz (Fig. 3.27).

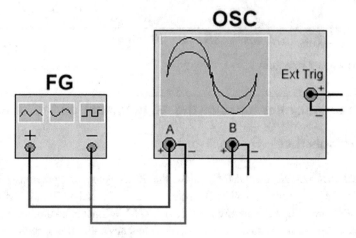

Fig. 3.26   Function generator FG is connected to oscilloscope

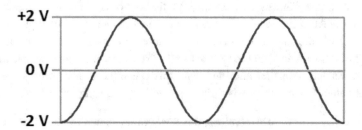

Fig. 3.27   Waveform generated by function generator FG in Fig. 3.26

Now, put the DMM in the AC voltmeter mode and connect it to the function generator (Fig. 3.28). Equivalent circuit for Fig. 3.28 is shown in Fig. 3.29. Since the internal resistance of AC voltmeter is much bigger than 50 Ω, all of the input voltage drops across it. Change the frequency to the values shown in Table 3.1 and fill the table (Don't change the amplitude knob, all of the measurements must be done with a sinusoidal signal with amplitude of 2 V). Note that the values shown by AC voltmeter decreases as frequency increases. You can use MATLAB to draw the graph of data shown in Table 3.1 as well (see Sect. A.7 in Appendix A).

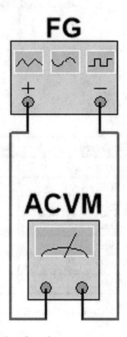

Fig. 3.28   AC voltmeter is connected to function generator

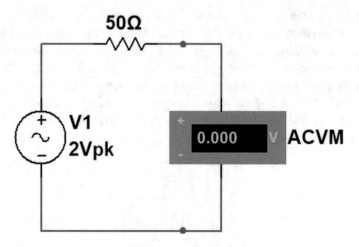

Fig. 3.29 Equivalent circuit for Fig. 3.28

Table 3.1 Readings of AC voltmeter for different frequency values

Frequency	Reading of AC volt meter
50 Hz	
100 Hz	
1 kHz	
5 kHz	
10 kHz	
15 kHz	
20 kHz	
30 kHz	
40 kHz	
50 kHz	
75 kHz	
100 kHz	

Reading of the AC voltmeter for low values of frequency is around $\frac{2}{\sqrt{2}} = \sqrt{2} = 1.41\,\text{V}$. Increase the frequency until you observe $\frac{1}{\sqrt{2}} \times \sqrt{2} = 1\,\text{V}$ on the display. This frequency is the cut-off frequency of the AC voltmeter. 2π times this frequency gives the value of ω_0 in Fig. 3.25 for AC voltmeter.

Now put the DMM in AC milli Amper measurement mode (Fig. 3.30). Equivalent circuit for Fig. 3.30 is shown in Fig. 3.31. The internal resistance of AC voltmeter is much smaller than 50 Ω. Therefore, the resistance seen by the function generator is around 50 Ω and a current with peak value of $\frac{2}{50} = 40\,\text{mA}$ pass through the milli Ammeter (Fig. 3.32). Change the frequency to the values shown in Table 3.2 and fill the table (Don't change the amplitude knob, all of the measurements must be done with a sinusoidal signal with amplitude of 2 V). Note that the values shown by AC milli Ammeter decreases as frequency increases. You can use MATLAB to draw the graph of data shown in Table 3.2 as well (see Sect. A.7 in Appendix A).

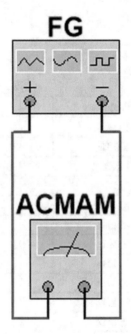

Fig. 3.30 AC milli ammeter is connected to the function generator

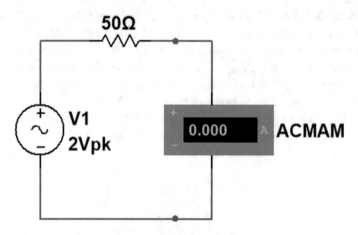

Fig. 3.31 Equivalent circuit for Fig. 3.30

Table 3.2 Readings of AC milli ammeter for different frequency values	Frequency			Reading of AC volt meter
	50 Hz			
	100 Hz			
	1 kHz			
	5 kHz			
	10 kHz			
	15 kHz			
	20 kHz			
	30 kHz			
	40 kHz			
	50 kHz			
	75 kHz			
	100 kHz			

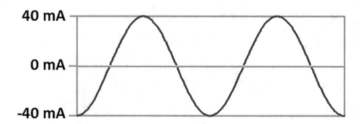

Fig. 3.32 Current waveform for Fig. 3.30

Reading of the AC milli Ammeter for low values of frequency is around $\frac{40\,mA}{\sqrt{2}} = 28.28\,mA$. Increase the frequency until you observe $\frac{1}{\sqrt{2}} \times 28.28\,mA = 20\,mA$ on the display. This frequency is the cut-off frequency of the AC milli Ammeter. 2π times this frequency gives the value of ω_0 in Fig. 3.25 for AC milli Ammeter.

Reference for Further Study

1. Asadi F., Eguchi K., Electronic Measurement: A Practical Approach, Springer, 2021.

Circuit Theorems

4

4.1 Introduction

This chapter studies the Kirchhoff's laws, Nodal/Mesh analysis, Thevenin's theorem and maximum power transfer theorem. You will see that aforementioned tools are really correct. This chapter contains 5 experiments.

4.2 Voltage Division, Current Division and KCL

4.2.1 Introduction

Consider the voltage divider circuit shown in Fig. 4.1. Using the basic circuit theory $V_1 = \frac{R_1}{R_1+R_2} V_{in}$ and $V_2 = \frac{R_2}{R_1+R_2} V_{in}$.

Fig. 4.1 Voltage divider circuit

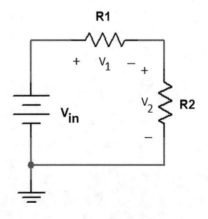

© The Author(s), under exclusive license to Springer Nature Switzerland AG 2023
F. Asadi, *Electric Circuits Laboratory Manual*, Synthesis Lectures
on Electrical Engineering, https://doi.org/10.1007/978-3-031-24552-7_4

Consider the current divider circuit shown in Fig. 4.2. Using the basic circuit theory $I_2 = \frac{R_3}{R_2+R_3}I_1$ and $I_3 = \frac{R_2}{R_2+R_3}I_1$.

Fig. 4.2 Current divider

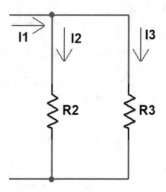

This experiment studies the voltage and current divider circuits.

4.2.2 Procedure

Prepare three resistors with values of 1 kΩ, 2.2 kΩ and 3.3 kΩ and measure their resistance with a DMM. Then make the circuit shown in Fig. 4.3. Connect the negative probe (black probe) of the DMM to the ground and measure the voltage of node A, B and C. Compare the measured voltages with values predicted by theory: $V_A = 5\,\text{V}$, $V_B = \frac{R_2+R_3}{R_1+R_2+R_3}V_A = 4.235\,\text{V}$ and $V_C = \frac{R_3}{R_1+R_2+R_3}V_A = 2.545\,\text{V}$.

Fig. 4.3 V1 voltage is divided between R1, R2 and R3

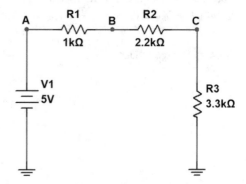

Now, make the circuit shown in Fig. 4.4 and measure the voltage of node A and B (with respect to ground). Compare the measured values with values predicted by theory: $V_A = 5V$, $V_B = \dfrac{\frac{R_2 \times R_3}{R_2 + R_3}}{R_1 + \frac{R_2 \times R_3}{R_2 + R_3}} V_A \approx 2.85V$. Calculate the $I1, I2$ and $I3$ with the aid of Ohm's law: $I1 = \dfrac{V_A - V_B}{R1}$, $I2 = \dfrac{V_B}{R2}$ and $I3 = \dfrac{V_B}{R3}$. Verify that $I1 = I2 + I3$, $I2 = \dfrac{R_3}{R_2 + R_3} I1$ and $I3 = \dfrac{R_2}{R_2 + R_3} I1$.

Fig. 4.4 I1 Current is divided between R2 and R3

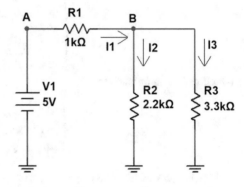

4.3   Nodal Analysis

4.3.1   Introduction

Nodal analysis is a Kirchhoff Current Law (KCL) based method and it gives the voltage of nodes to us. Let's review the nodal analysis with the aid of an example (Fig. 4.5).

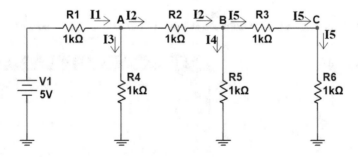

Fig. 4.5 Given sample circuit

According to the above circuit we have:

$$\begin{cases} I1 = I2 + I3 \\ I2 = I4 + I5 \\ I5 = I5 \end{cases} \tag{4.1}$$

This equals to

$$\begin{cases} \frac{5-V_A}{1} = \frac{V_A-V_B}{1} + \frac{V_A}{1} \\ \frac{V_A-V_B}{1} = \frac{V_B}{1} + \frac{V_B-V_C}{1} \\ \frac{V_B-V_C}{1} = \frac{V_C}{1} \end{cases} \tag{4.2}$$

which can be simplified to

$$\begin{cases} 3V_A - V_B + 0V_C = 5 \\ V_A - 3V_B + V_C = 0 \\ 0V_A + V_B - 2V_C = 0 \end{cases} \tag{4.3}$$

Matrix form of Eq. (4.3) is:

$$\begin{bmatrix} 3 & -1 & 0 \\ 1 & -3 & 1 \\ 0 & 1 & -2 \end{bmatrix} \begin{bmatrix} V_A \\ V_B \\ V_C \end{bmatrix} = \begin{bmatrix} 5 \\ 0 \\ 0 \end{bmatrix} \tag{4.4}$$

Therefore,

$$\begin{bmatrix} V_A \\ V_B \\ V_C \end{bmatrix} = \begin{bmatrix} 3 & -1 & 0 \\ 1 & -3 & 1 \\ 0 & 1 & -2 \end{bmatrix}^{-1} \times \begin{bmatrix} 5 \\ 0 \\ 0 \end{bmatrix}$$

This equation can be solved with the aid of MATLAB commands shown in Fig. 4.6.

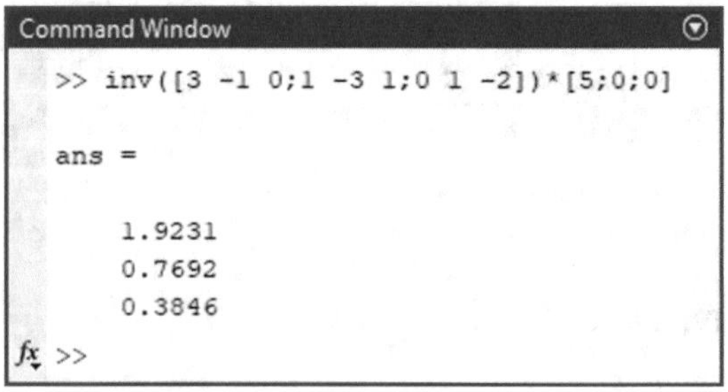

Fig. 4.6 MATLAB calculations

```
Command Window

>> inv([3 -1 0;1 -3 1;0 1 -2])*[5;0;0]

ans =

    1.9231
    0.7692
    0.3846

fx >>
```

According to Fig. 4.6, $\begin{bmatrix} V_A \\ V_B \\ V_C \end{bmatrix} = \begin{bmatrix} 1.6667 \\ 0 \\ -0.8333 \end{bmatrix}$.

4.3.2 Procedure

Make the circuit shown in Fig. 4.7. Measure the voltage of node A, B and C and write them in Table 4.1. Compare the measured values with the calculated values, i.e., $\begin{bmatrix} V_A \\ V_B \\ V_C \end{bmatrix} = \begin{bmatrix} 1.6667 \\ 0 \\ -0.8333 \end{bmatrix}$. Try to find the source of discrepancy between the two sets.

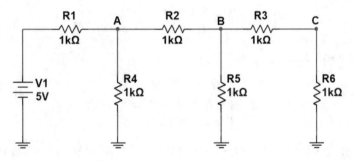

Fig. 4.7 Measure the potential of node A, B and C

Table 4.1 Voltage of node A, B and C

Node	Voltage
A	
B	
C	

4.4 Mesh Analysis

4.4.1 Introduction

Mesh analysis is a Kirchhoff Voltage Law (KVL) based method and it gives the current of meshes to us. Let's review the mesh analysis with the aid of an example.

According to Fig. 4.8 we have:

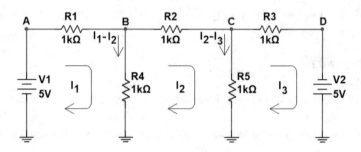

Fig. 4.8 Given sample circuit

$$\begin{cases} -5 + 1 \times I_1 + 1 \times (I_1 - I_2) = 0 \\ -(I_1 - I_2) + 1 \times I_2 + (I_2 - I_3) = 0 \\ -(I_2 - I_3) + 1 \times I_3 - 5 = 0 \end{cases} \qquad (4.5)$$

which can be written as:

$$\begin{cases} 2I_1 - I_2 + 0I_3 = 5 \\ -I_1 + 3I_2 - I_3 = 0 \\ 0I_1 - I_2 + 2I_3 = 5 \end{cases} \qquad (4.6)$$

This linear system can be solved with the aid of MATLAB. According to Fig. 4.9,
$$\begin{bmatrix} I_1 \\ I_2 \\ I_3 \end{bmatrix} = \begin{bmatrix} 3.7500 \\ 2.500 \\ 3.7500 \end{bmatrix}.$$

Fig. 4.9 MATLAB calculation

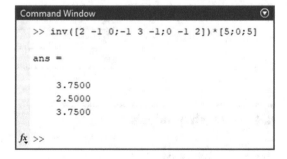

4.4.2 Procedure

Make the circuit shown in Fig. 4.10 and measure the voltage of nodes A, B, C and D. Write the measured voltages in Table 4.2.

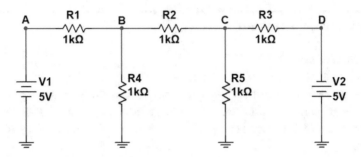

Fig. 4.10 Measure the potential of point A, B, C and D

Table 4.2 Voltage of node A, B, C and D	Node	Voltage
	A	
	B	
	C	
	D	

Use the voltages in Table 4.2 to calculate the values of currents shown in Fig. 4.11. $I_1 = \frac{V_A - V_B}{R1}$, $I_1 - I_2 = \frac{V_B}{R4}$, $I_2 = \frac{V_B - V_C}{R2}$, $I_2 - I_3 = \frac{V_C}{R5}$ and $I_3 = \frac{V_C - V_D}{R3}$. You can use an ammeter to measure the I_1, I_2 and I_3 instead of using Ohm's law.

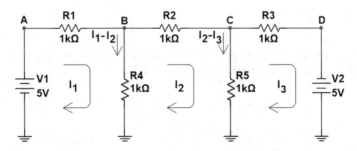

Fig. 4.11 I_1, I_2 and I_3 currents

Compare the measured values with the calculated values, i.e., $\begin{bmatrix} I_1 \\ I_2 \\ I_3 \end{bmatrix} = \begin{bmatrix} 3.7500 \\ 2.500 \\ 3.7500 \end{bmatrix}$.

Try to find source of discrepancy if any.

4.5 Thevenin Theorem

4.5.1 Introduction

Thevenin's theorem states that for any linear electrical network containing only voltage sources, current sources and resistances can be replaced at terminals A–B by an equivalent combination of a voltage source Vth in a series connection with a resistance Rth (Fig. 4.12).

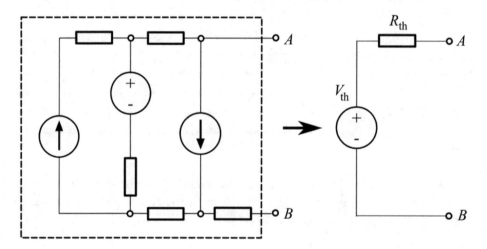

Fig. 4.12 Thevenin theorem

Let's review the Thevenin theorem with an example. Assume that we want to calculate the Thevenin equivalent circuit for terminals A–B shown in Fig. 4.13.

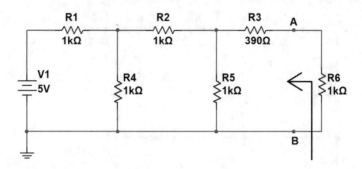

Fig. 4.13 Given sample circuit

We need to replace the V1 with short circuit (Fig. 4.14) in order to calculate the Rth. According to Fig. 4.14, the resistance seen is $390 + \frac{1.5 \times 1}{1.5 + 1} = 390 + 0.6 = 990 \, \Omega$.

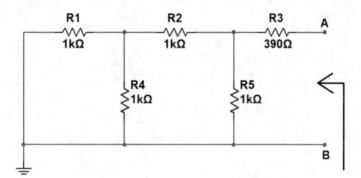

Fig. 4.14 Voltage source V1 is replaced with short circuit

Thevenin voltage equals to the open circuit voltage. According to Fig. 4.15, $V_{oc} = V_A - V_B = V_D - V_B = V_D$. It is quite simple to show that $V_D = 1V$. Therefore, $V_{th} = V_{oc} = 1V$.

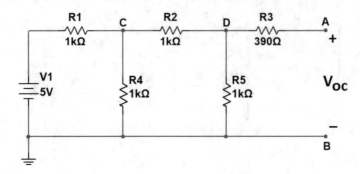

Fig. 4.15 Measurement of Thevenin voltage

Thevenin equivalent circuit seen from terminals A and B are shown in Fig. 4.16.

Fig. 4.16 Thevenin equivalent
circuit for Fig. 4.13

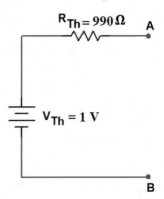

Voltage drop across R6 in Fig. 4.13 must be around $\frac{1}{1+0.99} \times 1 = 0.503$ V according
to Fig. 4.17.

Fig. 4.17 Addition of resistor
R6 to the Thevenin equivalent
circuit

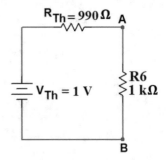

This experiment studies the Thevenin equivalent circuit.

4.5.2 Procedure

Make the circuit shown in Fig. 4.18. Measure the voltage drop across the R6 and compare it with the value calculated using hand analysis (0.503 V). Try to explain the reason of discrepancy if any.

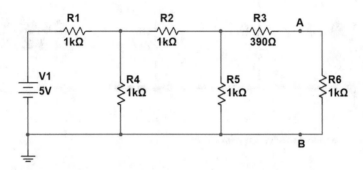

Fig. 4.18 Given circuit

We want to find the Thevenin equivalent circuit seen from points A and B. Remove the R6 and measure the open circuit voltage difference between points A and B (Fig. 4.19). Measured voltage gives the Vth to you.

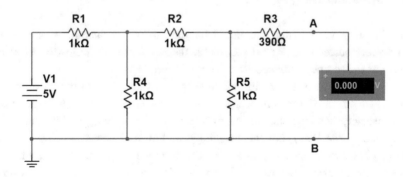

Fig. 4.19 Measurement of voltage difference between node A and B

Now use an ammeter to measure the short circuit current (Fig. 4.20). Rth equals to the ratio of open circuit voltage to the short circuit current.

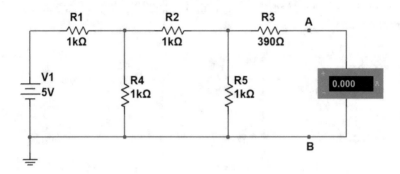

Fig. 4.20 Measurement of short circuit current

Compare the obtained Vth and Rth values with values obtained using hand analysis. Try to explain the reason of discrepancy if any.

4.6   Maximum Power Transfer

4.6.1   Introduction

The maximum power transfer theorem states that, to obtain maximum external power from a power source with internal resistance, the resistance of the load must equal the resistance of the source as viewed from its output terminals.

Note that the theorem results in maximum power transfer from the power source to the load, and not maximum efficiency. If the load resistance is made larger than the source resistance, then efficiency increases (since a higher percentage of the source power is transferred to the load), but the magnitude of the load power decreases (since the total circuit resistance increases). If the load resistance is made smaller than the source resistance, then efficiency decreases (since most of the power ends up being dissipated in the source). Although the total power dissipated increases (due to a lower total resistance), the amount dissipated in the load decreases.

The maximum power transfer theorem has a close relationship with Thevenin theorem as well. When you have a network, the maximum power is transferred to the load if the load equals to the Thevenin resistance seen from the load terminals.

This experiment studies the maximum power theorem.

4.6.2   Procedure

Make the circuit shown in Fig. 4.21. Use the values shown in Table 4.3 and fill it. Note that dissipated power in R6 can be calculated using the $\frac{V_{AB}^2}{R_6}$. The 960 Ω resistor in Table 4.3 can be made by series connection of a 470 Ω, 390 Ω and 100 Ω resistors (960 Ω = 470 Ω + 390 Ω + 100 Ω). When dissipated power in R6 has its maximum value? Compare this value with the Rth found in the previous experiment.

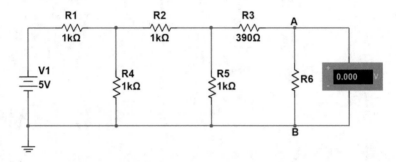

Fig. 4.21 Measurement of voltage difference between nodes A and B for different values of R6

Table 4.3 Suggested values for R6 and power dissipated in R6

Resistor R6 value	100 Ω	470 Ω	820 Ω	960 Ω	2.2 kΩ	4.7 kΩ	10 kΩ	22 kΩ
Voltmeter reading								
Dissipated power								

References for Further Study

1. Asadi F., Essential Circuit Analysis using Proteus, Springer, 2022.
2. Asadi F., Essential Circuit Analysis using LTspice, Springer, 2022.
3. Asadi F., Essential Circuit Analysis using NI Multisim and MATLAB, Springer, 2022.
4. Asadi F., Electric Circuit Analysis with EasyEDA, Springer, 2022.

First Order and Second Order Circuits

5

5.1   Introduction

In this chapter you will study the behavior of RC, RL and RLC circuits. This chapter contains 4 experiments.

5.2   Output Resistance of Function Generator

5.2.1   Introduction

A function generator has a limited non-zero output resistance similar to any other source. This experiment shows how the output resistance of a function generator can be measured. Most of the function generators have 50 Ω output resistance.

5.2.2   Procedure

Connect the oscilloscope to the function generator (Fig. 5.1) and generate a sinusoidal waveform with peak value of 2 V and frequency of 1 kHz (Fig. 5.2). Equivalent circuit for Fig. 5.1 is shown in Fig. 5.3. The input resistance of the oscilloscope (RinScope) is much bigger than the output resistance of the function generator (RoFG). Therefore, all of the input voltage drops across the input resistance of the oscilloscope. Write the peak value of the voltage that you see on the scope screen and call it $V_{p,oc}$.

© The Author(s), under exclusive license to Springer Nature Switzerland AG 2023   77
F. Asadi, *Electric Circuits Laboratory Manual*, Synthesis Lectures
on Electrical Engineering, https://doi.org/10.1007/978-3-031-24552-7_5

Fig. 5.1 Function generator is
connected to the oscilloscope

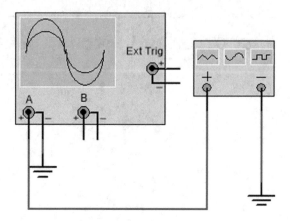

Fig. 5.2 Waveform generated
by function generator

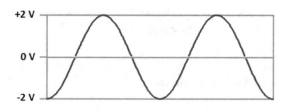

Fig. 5.3 Equivalent circuit for
Fig. 5.1

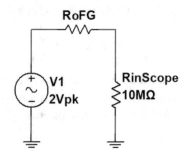

Now, connect a 56 Ω resistor to output of the function generator (Fig. 5.4) and measure
the peak of the waveform shown on the oscilloscope screen. Call this value $V_{p,loaded}$.
Equivalent circuit for Fig. 5.4 is shown in Fig. 5.5. Equivalent resistance for parallel
connection of a 56 Ω and a 10 MΩ resistors is 56 Ω. Therefore we can write $\frac{56}{RoFG+56} \times$
$V_{p,oc} = V_{p,loaded} \Rightarrow RoFG = (\frac{V_{p,oc}}{V_{p,loaded}} - 1) \times 56$. Use this formula to calculate the
output resistance of function generator. Generally, the output impedance of the function
generators is around 50 Ω.

Fig. 5.4 Connecting a 56 Ω
resistor to output of signal
generator

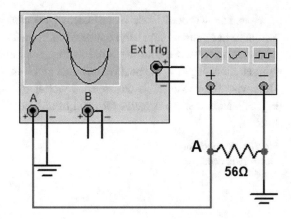

Fig. 5.5 Equivalent circuit for
Fig. 5.4

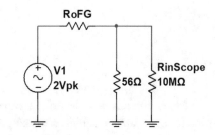

5.3   Step Response of RC Circuit

5.3.1   Introduction

Capacitance of an electrolytic capacitor and maximum voltage that can be applied to it
are written on the capacitor directly. For instance, in Fig. 5.6, an electrolytic capacitor
with value of 2200 µF and maximum working voltage of 16 V is shown.

Fig. 5.6 An electrolytic
capacitor

Note that terminal leads of an electrolytic capacitor are not the same: One of the terminals has a minus sign behind it (see Fig. 5.6). This is terminal is the negative terminal of the capacitor and it must be connected to lower potential. The other terminal (terminal without minus sign) is the positive terminal and it must be connected to positive potential. Otherwise the capacitor is destroyed. Symbol of an electrolytic capacitor is shown in Fig. 5.7. Note that the negative terminal of an electrolytic capacitor is shown with curved line.

Fig. 5.7 Symbol for
electrolytic capacitor

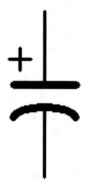

If you connect two electrolytic capacitors in series, you will obtain a capacitor with lower capacitance but higher working voltage. For instance, assume that we connected two electrolytic capacitors with value of 1 μF and working voltage of 50 V together as shown in Fig. 5.8. Equivalent capacitor for this series connection is 0.5 μF and 100 V.

Fig. 5.8 Series connected
electrolytic capacitors

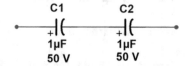

If you connect two electrolytic capacitors in parallel, you will obtain a capacitor with higher capacitance but working voltage doesn't change. For instance, assume that we connected two electrolytic capacitors with value of 1 μF and working voltage of 50 V together as shown in Fig. 5.9. Equivalent capacitor for this series connection is 2 μF and 50 V.

Fig. 5.9 Parallel connected
electrolytic capacitors

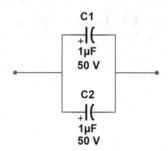

In ceramic capacitors, value of the capacitor is shown indirectly. Third number (from left) shows the number of zeros that must be put in front of the first two numbers (obtained number is in pF). For instance, in Fig. 5.10, 103 is printed on the capacitor. This means 10,000 pF or 10 nF. Terminal leads of ceramic capacitors are the same. You can connect the higher/lower potential to either of them. Ceramic capacitors have working voltage of at least 50 V.

Fig. 5.10 Ceramic capacitor

This experiment studies the step response (i.e., output for unit step input) of first order RC circuit shown in Fig. 5.11. Unit step input is shown in Fig. 5.12. MATLAB's step command calculates the step response with zero initial conditions. Note that the 50 Ω resistor in Fig. 5.11 shows the output resistance of the signal generator.

Fig. 5.11 RC circuit with step input

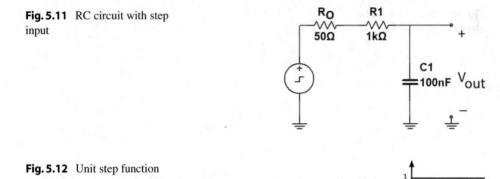

Fig. 5.12 Unit step function

Step response of the circuit can be calculated using the commands shown in Fig. 5.13. Output of this code is shown in Fig. 5.14.

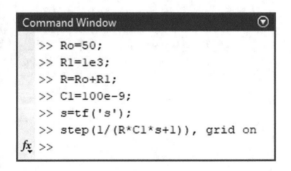

```
Command Window
    >> Ro=50;
    >> R1=1e3;
    >> R=Ro+R1;
    >> C1=100e-9;
    >> s=tf('s');
    >> step(1/(R*C1*s+1)), grid on
fx >>
```

Fig. 5.13 MATLAB code

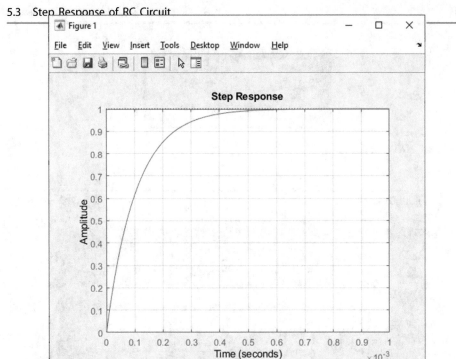

Fig. 5.14 Output of MATLAB code

Write the KVL for circuit shown in Fig. 5.11 and use the Laplace transform to solve the obtained differential equation in order to ensure that result shown in Fig. 5.14 is correct.

Assume that we want to obtain the output of the circuit shown in Fig. 5.11 for signal shown in Fig. 5.15. This time we need to use the commands shown in Fig. 5.16. Output of this code is shown in Fig. 5.17. Note that final value of capacitor voltage is 5 V.

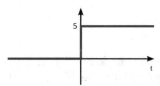

Fig. 5.15 Step function with amplitude of 5

```
Command Window                                    ⊙
  >> Ro=50;
  >> R1=1e3;
  >> R=Ro+R1;
  >> C1=100e-9;
  >> step(5*1/(R*C1*s+1)), grid on
fx >> |
```

Fig. 5.16 MATLAB code

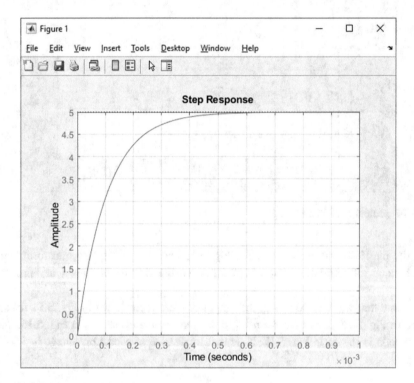

Fig. 5.17 Output of MATLAB code

   You can read a point by clicking on it. For instance, according to Fig. 5.18, voltage of capacitor reaches 3.15 V after 0.000104 s = 104 μs.

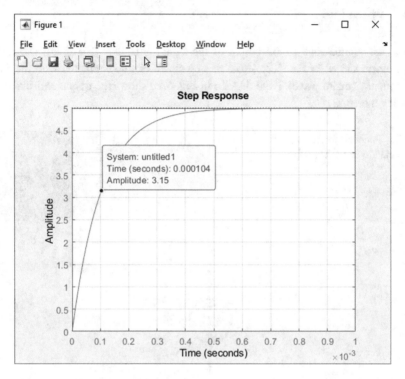

Fig. 5.18   Output is 3.15 at t = 0.000104 s

Remember that capacitor in first order RC circuit is charged up to 63% of source within one time constant. Source voltage is 5 V and time constant is 105 μs (Fig. 5.19). Therefore, we expect the voltage to reach $0.63 \times 5 = 3.15$ V within 105 μs. Figure 5.18 proves this.

Fig. 5.19   MATLAB calculations

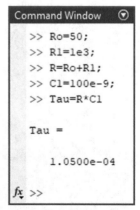

5.3.2 Procedure

Connect the oscilloscope to the output of the signal generator (Fig. 5.20) and generate the waveform shown in Fig. 5.21 (Role of step signal can be simulated with the aid of a low frequency square wave). Note that Ro in Fig. 5.20 shows the output resistance of the function generator.

Fig. 5.20 Function generator is connected to an oscilloscope

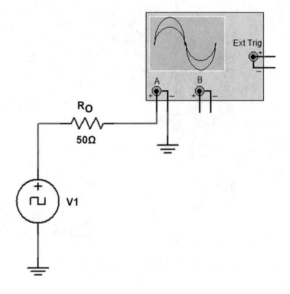

Fig. 5.21 Square waveform generated by signal generator

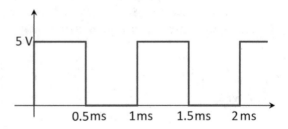

Connect the function generator to the series connection of the R1 and C1 and observe the capacitor voltage (Fig. 5.22). Compare the waveform on the oscilloscope screen with Fig. 5.18. Measure the time required to go from $0V$ to $0.63 \times 5 = 3.15V$ and compare it with the time constant $(\tau = (R_O + R_1) \times C_1 = 105\mu s)$ of the circuit.

Fig. 5.22 Square wave is applied to the RC circuit

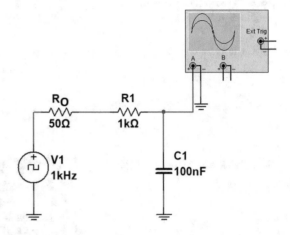

5.4 Step Response of RL Circuit

5.4.1 Introduction

Step response of the circuit shown in Fig. 5.23 is studied in this experiment. 50 Ω resistance shows the output resistance of the signal generator.

Fig. 5.23 RL circuit with step input

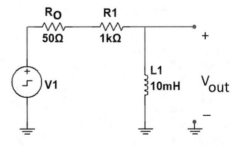

Output of the circuit for input shown in Fig. 5.24 can be calculated using the commands shown in Fig. 5.25 (All of the initial conditions are assumed to be zero). Output of this code is shown in Fig. 5.26.

Fig. 5.24 Step function with amplitude of 5

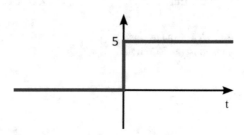

Fig. 5.25 MATLAB code

```
Command Window                                      ⊙
   >> Ro=50;
   >> Rl=le3;
   >> R=Ro+Rl;
   >> Ll=10e-3;
   >> s=tf('s');
   >> step(5*Ll*s/(R+Ll*s)), grid on
fx >>
```

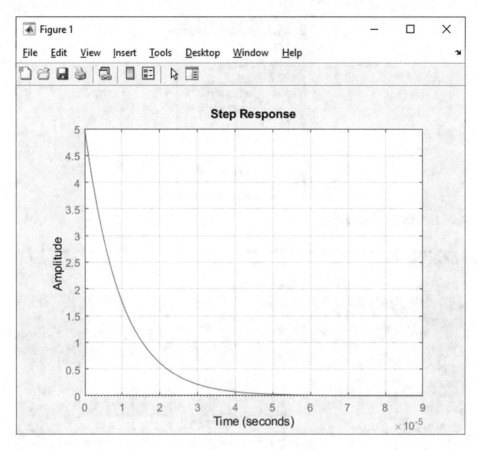

Fig. 5.26   Output of MATLAB code

Write the KVL for circuit shown in Fig. 5.23 and use the Laplace transform to solve the obtained differential equation in order to ensure that result shown in Fig. 5.26 is correct.

According to Fig. 5.27, value of time constant for this circuit is 9.5238 μs. We expect the curve reach to 37% of its initial value after one time constant. Let's test it. Figure 5.28 shows that after 9.47 μs the curve reaches to $0.37 \times 5 = 1.85$ V.

Fig. 5.27 MATLAB calculations

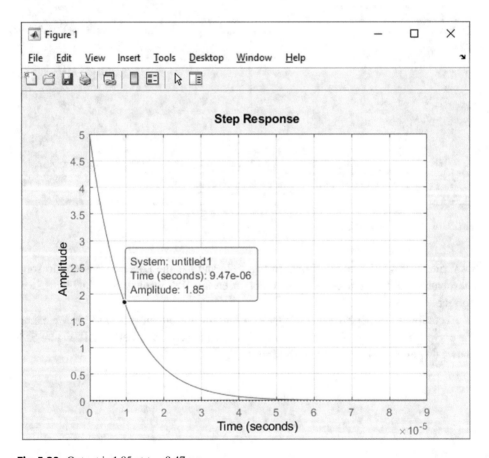

Fig. 5.28 Output is 1.85 at t = 9.47 μs

5.4.2   Procedure

Connect the oscilloscope to the output of the signal generator (Fig. 5.29) and generate the waveform shown in Fig. 5.30. Note that Ro in Fig. 5.29 shows the output resistance of the function generator.

Fig. 5.29 Function generator is connected to an oscilloscope

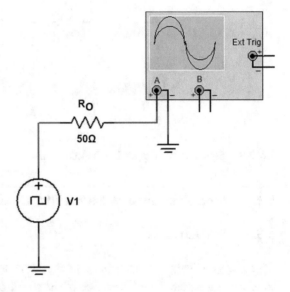

Fig. 5.30 Square waveform generated by signal generator

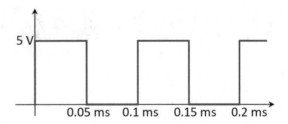

Connect the function generator to the series connection of the R1 and L1 and observe the inductor voltage (Fig. 5.31). Compare the waveform on the oscilloscope screen with Fig. 5.28. Measure the time required to go from $5V$ to $5 - 0.63 \times 5 = 1.85V$ and compare it with the time constant ($\tau = \frac{L_1}{R_O + R_1}$) of the circuit.

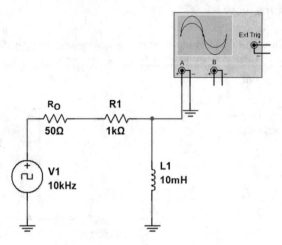

Fig. 5.31 Square wave is applied to the RC circuit

5.5   Step Response of Series RLC Circuit

5.5.1   Introduction

Step response of the circuit shown in Fig. 5.32 is studied in this experiment. 50 Ω resistance shows the output resistance of the signal generator.

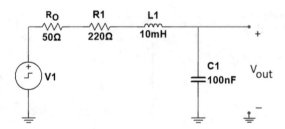

Fig. 5.32 RLC circuit with step input

Output of the circuit for input shown in Fig. 5.33 can be calculated using the commands shown in Fig. 5.34 (All of the initial conditions are assumed to be zero). Output of this code is shown in Fig. 5.35. The response shown in Fig. 5.35 is underdamped. Response is critically damped for $R1 = 2\sqrt{\frac{L}{C}} = 632.455\ \Omega$. Write the KVL for circuit shown in Fig. 5.32 and use the Laplace transform to solve the obtained differential equation in order to ensure that result shown in Fig. 5.35 is correct.

Fig. 5.33 Step function with amplitude of 5

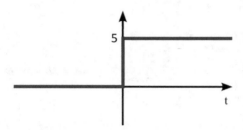

Fig. 5.34 MATLAB code

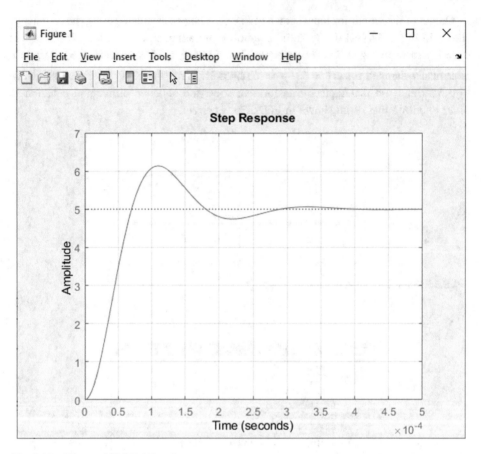

Fig. 5.35 Output of MATLAB code

Let's study the behavior of the circuit for R1 = 680 Ω as well. Input signal is the same as Fig. 5.33. The code shown in Fig. 5.36 analyze the behavior of the circuit for R1 = 680 Ω. Output of this code is shown in Fig. 5.37. The response shown in Fig. 5.37 is overdamped.

Fig. 5.36 MATLAB code

```
Command Window                                    ⊙
  >> Ro=50;
  >> R1=680;
  >> R=Ro+R1;
  >> L1=10e-3;
  >> C1=100e-9;
  >> step(5*1/(L1*C1*s^2+R*C1*s+1)), grid on
fx >>
```

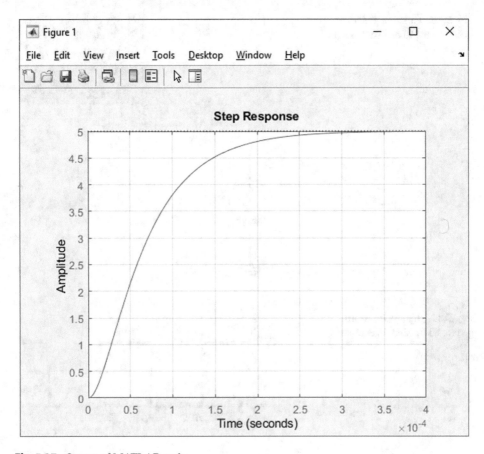

Fig. 5.37 Output of MATLAB code

5.5.2 Procedure

Connect the oscilloscope to the output of the signal generator (Fig. 5.38) and generate the waveform shown in Fig. 5.39. Note that Ro in Fig. 5.38 shows the output resistance of the function generator.

Fig. 5.38 Function generator
is connected to an oscilloscope

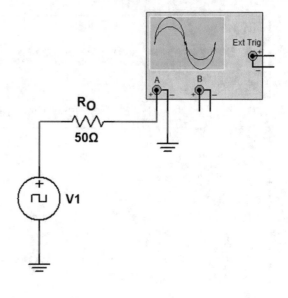

Fig. 5.39 Square waveform
generated by signal generator

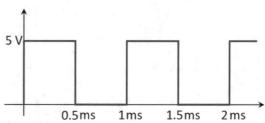

Connect the function generator to the series connection of the R1, L1 and C1 and observe the capacitor voltage (Fig. 5.40). Compare the waveform on the oscilloscope screen with Fig. 5.35. For instance, measure the maximum of the waveform and compare it with maximum of Fig. 5.35. Try to find the reason if there is a discrepancy.

Fig. 5.40 Square wave is
applied to the RC circuit

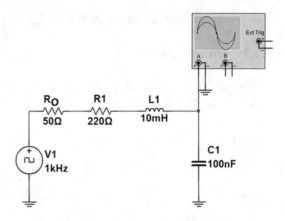

Change the resistor to 680 Ω (Fig. 5.41) and repeat the experiment. Compare the
obtained response with Fig. 5.37. Try to find the reason if there is a discrepancy.

Fig. 5.41 R1 is changed to
680 Ω

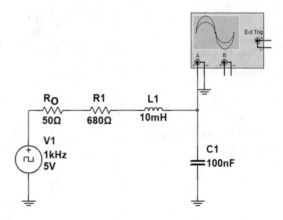

5.6   Step Response of Parallel RLC Circuit

5.6.1   Introduction

Step response of a parallel RLC circuit is studied in this experiment. Consider the circuit shown in Fig. 5.42. V1 shows a step voltage with amplitude of 5 V (Fig. 5.43).

Fig. 5.42 Parallel RLC circuit

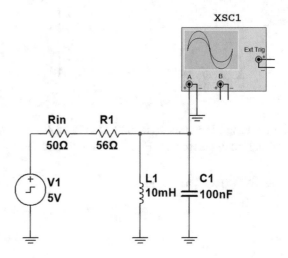

Fig. 5.43 Step function with amplitude of 5

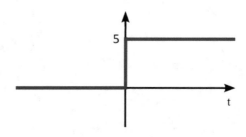

Using the source transformation theorems, the voltage source V1 and resistors Rin and R1 can be replaced with a current source with value of $\frac{5}{50+56} = 47.2\,\mathrm{mA}$ and a parallel resistor with value of $50 + 56 = 106\,\Omega$ (Fig. 5.44).

Fig. 5.44 Equivalent circuit for Fig. 5.42

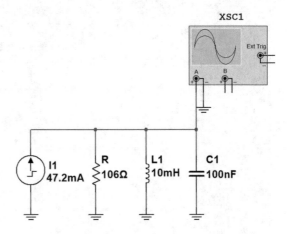

Let's use MATLAB to obtain the step response of the circuit shown in Fig. 5.44. MATLAB code shown in Fig. 5.45 draws the waveform of capacitor voltage. Output of this code is shown in Fig. 5.46.

```
Command Window

>> R=106;
>> L1=10e-3;
>> C1=100e-9;
>> s=tf('s');
>> Z=(1/R+1/L1/s+C1*s)^-1;
>> step(47.2e-3*Z), grid on
fx >>
```

Fig. 5.45 MATLAB code (Z is the impedance seen by the I1)

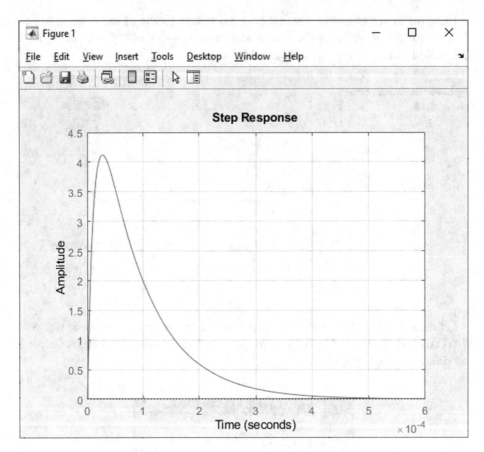

Fig. 5.46 Output of MATLAB code

5.6.2 Procedure

Connect the oscilloscope to the output of the signal generator (Fig. 5.47) and generate
the waveform shown in Fig. 5.48. Note that Ro in Fig. 5.47 shows the output resistance
of the function generator.

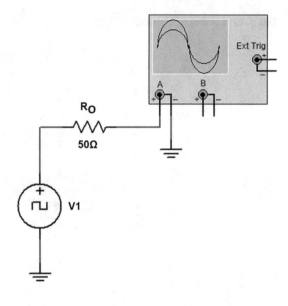

Fig. 5.47 Function generator is connected to an oscilloscope

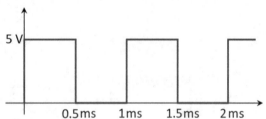

Fig. 5.48 Square waveform generated by signal generator

Connect the function generator to the rest of the circuit (Fig. 5.49). Equivalent circuit for Fig. 5.49 is shown in Fig. 5.50. Observe the capacitor voltage and compare the waveform on the oscilloscope screen with Fig. 5.46. For instance, measure the maximum of the waveform and compare it with maximum of Fig. 5.46. Try to find the reason if there is a discrepancy.

Fig. 5.49 Parallel RLC circuit

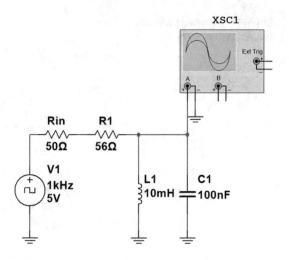

Fig. 5.50 Equivalent circuit
for parallel RLC circuit

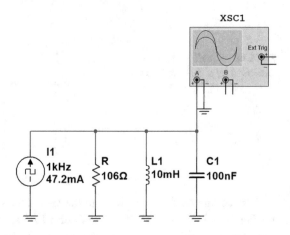

References for Further Study

1. Asadi F., Essential Circuit Analysis using Proteus, Springer, 2022.
2. Asadi F., Essential Circuit Analysis using LTspice, Springer, 2022.
3. Asadi F., Essential Circuit Analysis using NI Multisim and MATLAB, Springer, 2022.
4. Asadi F., Electric Circuit Analysis with EasyEDA, Springer, 2022.

Steady State DC and AC Analysis and Filters

6

6.1 Introduction

This chapter studies the DC and AC steady state behavior of electric circuits. Frequency response of filters are studies as well. This chapter contains 5 experiments.

6.2 Steady State DC Analysis

6.2.1 Introduction

Steady state behavior of inductor and capacitor in DC circuits is studied in this experiment. In DC steady state, inductors act like short circuit and capacitors act like open circuit.

Let's study an example. Consider the circuit shown in Fig. 6.1.

Fig. 6.1 Sample circuit

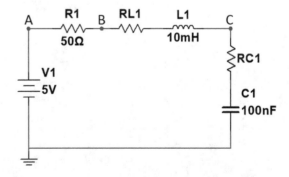

© The Author(s), under exclusive license to Springer Nature Switzerland AG 2023
F. Asadi, *Electric Circuits Laboratory Manual*, Synthesis Lectures
on Electrical Engineering, https://doi.org/10.1007/978-3-031-24552-7_6

DC steady state equivalent of Fig. 6.1 is shown in Fig. 6.2. Therefore, in DC steady state, the current drawn from the source will be 0 and potential of point A, B and C will be 5 V.

Fig. 6.2 Equivalent DC steady state model for Fig. 6.1

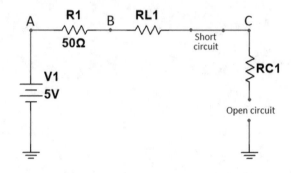

6.2.2   Procedure

Make the circuit shown in Fig. 6.3. Measure the voltage of points A, B and C. Voltage of node C gives the steady state capacitor voltage. Steady state current through the inductor can be calculated using $I_L = \frac{V_A - V_B}{R1}$. Compare the steady state current/voltage values with values predicted by theory.

Fig. 6.3 Simple RLC circuit

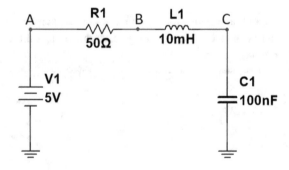

6.3 Steady State AC (Sinusoidal) Analysis

6.3.1 Introduction

Impedance of inductors and capacitors equals to $jL\omega$ and $\frac{-j}{C\omega}$, respectively. Let's review the steady state sinusoidal analysis with two examples.

Consider the circuit shown in Fig. 6.4 as our first example. RO shows the output resistance of the signal generator. We want to measure the phase difference between voltage of node B and A. Resistor R1 and capacitor C1 make a simple voltage divider circuit. So, $V_B = \frac{\frac{-j}{C_1\omega}}{R_1+\frac{-j}{C_1\omega}}V_A$. For instance, for $\omega = 2\pi \times 1000 = 6283.2\frac{\text{Rad}}{\text{s}}$ and values shown in Fig. 6.4, $V_B = \frac{\frac{-j}{C_1\omega}}{R_1+\frac{-j}{C_1\omega}}V_A = \frac{-j1.5915k}{1k-j1.5915k}V_A = 0.8467e^{-j32.143°}V_A$. Therefore, peak value of voltage of node B equals to 0.8467 times peak value of voltage of node A. Voltage of node B lags voltage of node A by 32.143°.

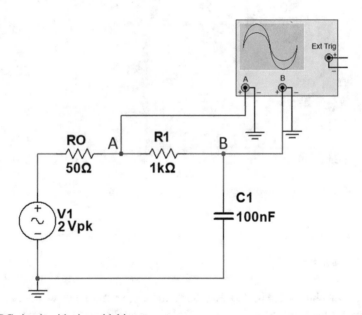

Fig. 6.4 RC circuit with sinusoidal input

Consider the circuit shown in Fig. 6.5 as our second example. Again, RO shows the output resistance of the signal generator. We want to measure the phase difference between voltage of node B and A. Resistor R1 and inductor L1 make a simple voltage divider circuit. So, $V_B = \frac{jL_1\omega}{R_1 + jL_1\omega}V_A$. For instance, for $\omega = 2\pi \times 1000 = 6283.2\frac{\text{Rad}}{\text{s}}$ and values shown in Fig. 6.5, $V_B = \frac{jL_1\omega}{R_1 + jL_1\omega}V_A = \frac{j62.832}{1k + j62.832}V_A = 0.0627e^{j86.4351°}V_A$. Therefore, peak value of voltage of node B equals to 0.0627 times peak value of voltage of node A. Voltage of node B leads voltage of node A by 86.4351°.

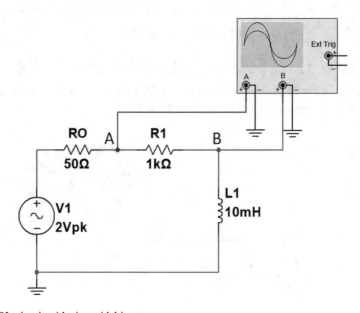

Fig. 6.5 RL circuit with sinusoidal input

6.3.2   Procedure

Make the circuit shown in Fig. 6.6. Ro shows the output resistance of the function generator. Generate the frequencies shown in Table 6.1 and fill the table. Amplitude of input voltage is 2 V in all of the measurements.

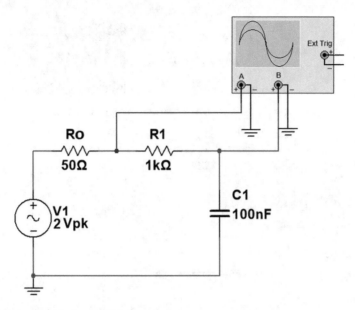

Fig. 6.6 RC circuit with sinusoidal input

Table 6.1 Steady state behavior of RC circuit shown in Fig. 6.6

Frequency	1 kHz	2 kHz	3 kHz	4 kHz	5 kHz
Peak value of Channel B					
Phase difference between Channel B and A					

Now change the circuit to what shown in Fig. 6.7 and fill the Table 6.2.

Compare the measured values (Tables 6.1 and 6.2) with values predicted by theory.

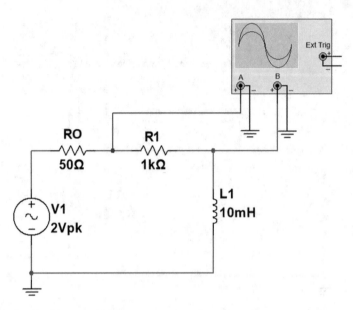

Fig. 6.7 RL circuit with sinusoidal input

Table 6.2 Steady state behavior of RL circuit shown in Fig. 6.7

Frequency	5 kHz	10 kHz	15 kHz	20 kHz	25 kHz
Peak value of Channel B					
Phase difference between Channel B and A					

6.4   Series and Parallel Resonance

6.4.1   Introduction

In this experiment resonance of series and parallel RLC circuits are studied. Let's start with the series RLC circuit. Consider the series RLC circuit shown in Fig. 6.8. 50 Ω resistor shows the output resistance of the function generator.

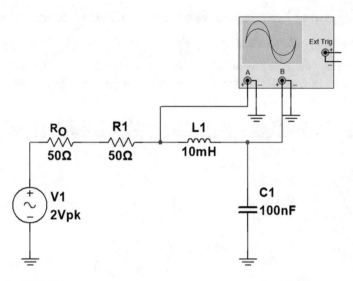

Fig. 6.8   Series RLC circuit

The impedance seen by input source V1 in Fig. 6.8 is $R_o + R_1 + jL_1\omega + \frac{-j}{C_1\omega}$. Magnitude of $R_o + R_1 + jL_1\omega + \frac{-j}{C_1\omega}$ equals to $\left| R_o + R_1 + jL_1\omega + \frac{-j}{C_1\omega} \right| = \sqrt{(R_o + R_1)^2 + (L_1\omega - \frac{1}{C_1\omega})^2}$. When $\omega = \frac{1}{\sqrt{L_1C_1}}$, $jL_1\omega + \frac{-j}{C_1\omega} = 0$ and the impedance seen by input source is purely resistive with value of $R_o + R_1$. At $\omega = \frac{1}{\sqrt{L_1C_1}}$ the impedance has minimum value, therefore the current drawn from the source is maximum. Resonance frequency of circuit shown in Fig. 6.8 is 5.0329 kHz (Fig. 6.9).

```
Command Window                                    ⊙

   >> L1=10e-3;
   >> C1=100e-9;
   >> fres=1/(2*pi*sqrt(L1*C1))

   fres =

        5.0329e+03

fx >>
```

Fig. 6.9   MATLAB calculations

Let's measure the capacitor's peak voltage at $f = 5.0329$ kHz. Capacitor's peak voltage can be calculated with the aid of $\left| \dfrac{\frac{-j}{C_1\omega}}{R_o+R_1+jL_1\omega-\frac{j}{C_1\omega}} \right| V_1 = \left| \dfrac{\frac{-j}{C_1\omega}}{R_o+R_1} \right| V_1 = \dfrac{1}{(R_o+R_1)C_1\omega} V_1$.
Capacitor's peak voltage is around 6.3246 V according to Fig. 6.10. Note that at resonant frequency, peak of capacitor (or inductor) voltage is bigger than peak of input source.

```
Command Window                                                    ⊙

  >> Vpeak=2;
  >> Ro=50;
  >> R1=50;
  >> R=Ro+R1;
  >> L1=10e-3;
  >> C1=100e-9;
  >> fres=1/2/pi/sqrt(L1*C1);
  >> w=2*pi*fres;
  >> XC=-j/w/C1;
  >> XL=j*L1*w;
  >> CapacitorPeakVoltage=abs(XC/(R+XL+XC)*Vpeak)

  CapacitorPeakVoltage =

        6.3246

fx >>
```

Fig. 6.10 MATLAB calculations

Now consider the parallel resonant circuit shown in Fig. 6.11. RL1 shows the resistance of the inductor. RL1 is assumed to be 0.1 Ω.

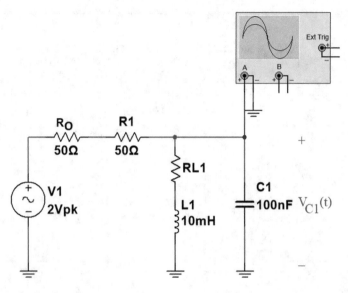

Fig. 6.11 Parallel RLC circuit

The MATLAB code shown in Fig. 6.12 draws the frequency response of $\frac{V_{C_1}(j\omega)}{V_1(j\omega)}$. Output of this code is shown in Fig. 6.13. Maximum of Magnitude (dB) graph occurs at the resonant frequency. The Magnitude (dB) graph is used to determine the resonant frequency.

Fig. 6.12 MATLAB code

```
Command Window
  >> Ro=50;
  >> R1=50;
  >> R=Ro+R1;
  >> L1=10e-3;
  >> RL1=0.1;
  >> C1=100e-9;
  >> s=tf('s');
  >> ZC1=1/C1/s;
  >> ZL1=L1*s;
  >> Z0=ZC1*(RL1+ZL1)/(ZC1+(RL1+ZL1));
  >> bode(Z0/(R+Z0)), grid on
fx >>
```

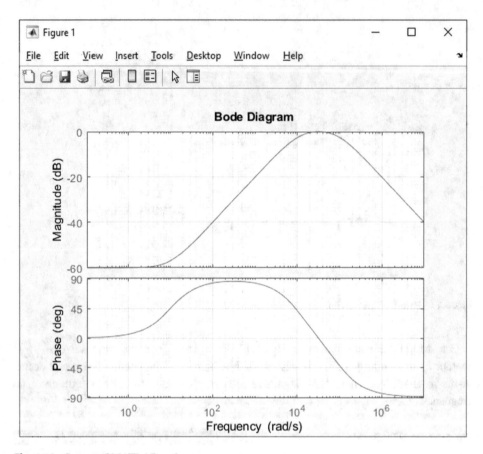

Fig. 6.13 Output of MATLAB code

Note that in Fig. 6.13, the horizontal axis is in Rad/s. Let's change it into Hz. Right click on the graph and click the Properties (Fig. 6.14). This opens the window shown in Fig. 6.15.

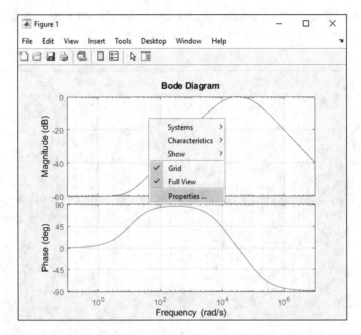

Fig. 6.14 A menu appears when you right click on the graph

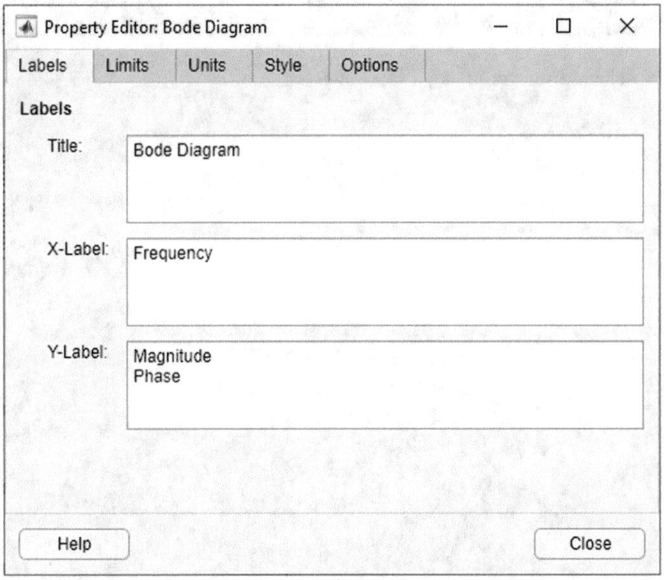

Fig. 6.15 Property Editor window

Go to the Units tab and change the Frequency to Hz and click the Close button
(Fig. 6.16). Now the horizontal axis is in Hz (Fig. 6.17).

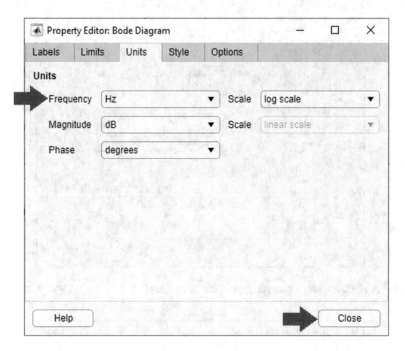

Fig. 6.16 Hz is selected for Frequency drop down list

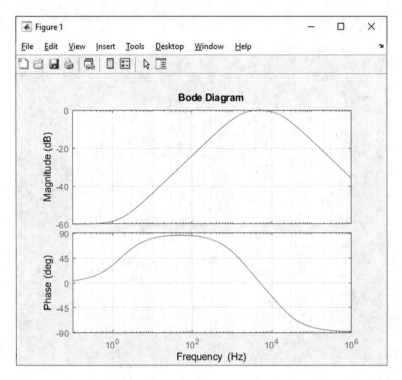

Fig. 6.17 Horizontal axis has unit of Hz

Add a cursor to the magnitude graph by clicking on it. Move the cursor to find the maximum of the graph. Maximum of magnitude graph occurs at 5.04 kHz with value of -0.00089 dB according to Fig. 6.18. -0.00089 dB equals to gain of 0.9999 (Fig. 6.19). Therefore, $\left| \frac{V_{C_1}(j2\pi \times 5040)}{V_1(j2\pi \times 5040)} \right| = 0.9999$.

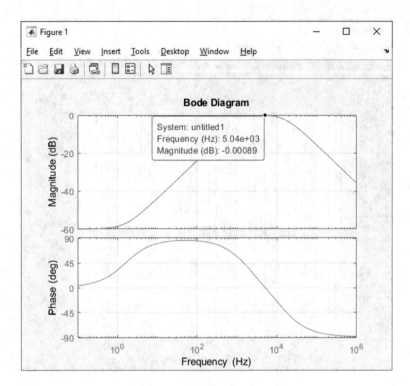

Fig. 6.18 Finding the maximum of Magnitude (dB) graph

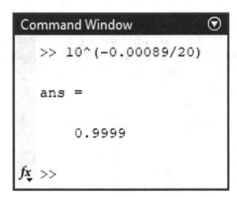

Fig. 6.19 Conversion of −0.00089 dB to normal gain

Figure 6.20 shows the frequency response of $\frac{V_{C_1}(j\omega)}{V_1(j\omega)}$ for RL1 = 1 Ω. This time the maximum occurs at 5.07 kHz with value of -0.00865 dB. -0.00865 dB equals to gain of $10^{\frac{-0.00865}{20}} = 0.9990$. Therefore, in this case $\left| \frac{V_{C_1}(j2\pi \times 5070)}{V_1(j2\pi \times 5070)} \right| = 0.9990$.

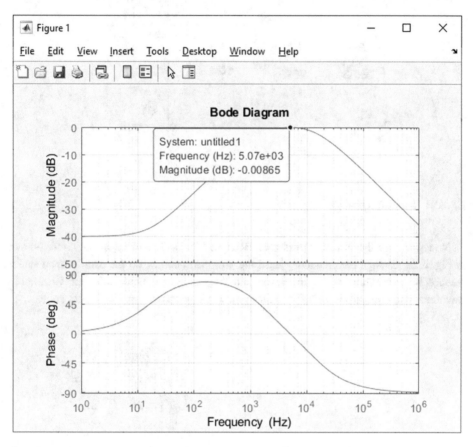

Fig. 6.20 Frequency response for RL1 = 1 Ω

6.4.2 Procedure

Make the circuit shown in Fig. 6.21. Note that Ro shows the output resistance of the function generator V1. Change the frequency of the V1 until Channel A waveform reaches its minimum and Channel B waveform reaches its maximum. Write the value of frequency. This is the resonance frequency of the circuit. Measure the peak value of capacitor voltage and compare it with the value calculated in Fig. 6.10.

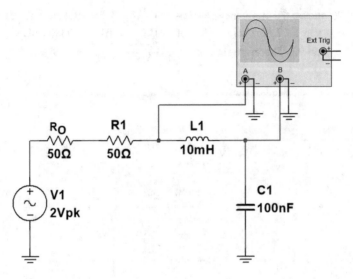

Fig. 6.21 Series RLC circuit

Now, measure the inductor resistance (RL1 in Fig. 6.11) and make the circuit shown in Fig. 6.22. Change the frequency until the waveform shown on the oscilloscope screen reaches its maximum. Write the resonance frequency and peak value of Channel A waveform and compare it with the values predicted by theory, i.e. MATLAB code.

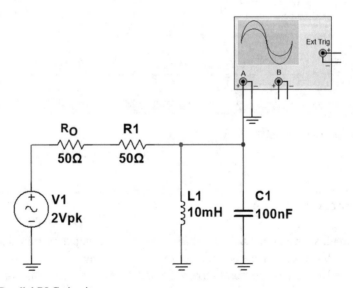

Fig. 6.22 Parallel RLC circuit

6.5   Low-Pass Filter

6.5.1   Introduction

A low-pass filter is a circuit that passes signals with a frequency lower than a selected cutoff frequency and attenuates signals with frequencies higher than the cutoff frequency.

In analysis of filter circuits Laplace domain (s domain) circuit analysis is a valuable tool. It permits us to obtain the transfer function ($\frac{\mathcal{L}(output)}{\mathcal{L}(input)}$) of the filter.

Let's study an example. The circuit shown in Fig. 6.23 is a low-pass filter (RO shows the output resistance of the signal generator). In Laplace domain circuit analysis, capacitor C is replaced with impedance of $\frac{1}{Cs}$. Therefore, the transfer function of circuit shown in

Fig. 6.23 will be $V_B(s) = \frac{\frac{1}{C_1 s}}{R_1 + \frac{1}{C_1 s}} V_A(s) = \frac{1}{R_1 C_1 s + 1} V_A(s) = \frac{1}{0.0001 s + 1} V_A(s) \Rightarrow \frac{V_B(s)}{V_A(s)} =$ $\frac{1}{0.0001 s + 1}$.

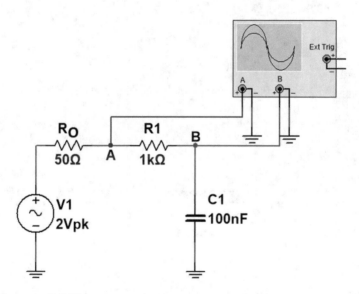

Fig. 6.23 Low-pass RC filter

Frequency response of $\frac{V_B(s)}{V_A(s)} = \frac{1}{0.0001 s + 1}$ can be drawn with the aid of the code shown in Fig. 6.24. Output of this code is shown in Fig. 6.25. Note that the Magnitude (dB) graph decreases as frequency increases. This shows that the filter is low-pass.

```
Command Window                          ⊙
>> s=tf('s');
>> H=1/(0.0001*s+1);
>> bode(H)
>> grid on
fx >>
```

Fig. 6.24 MATLAB code

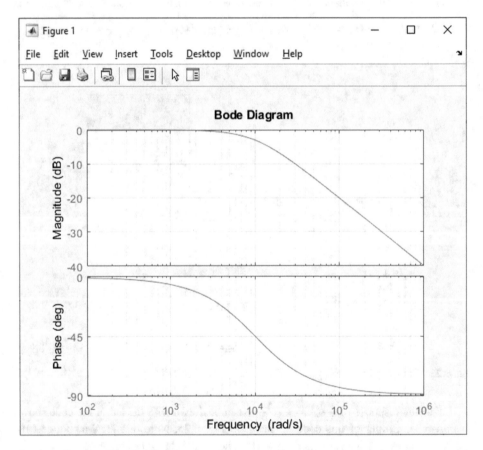

Fig. 6.25 Output of MATLAB code

Cut-off frequency is defined as the frequency which pass band gain decrease by 3 dB. According to Fig. 6.26, the pass band gain is around 0 dB. You can calculate the pass band gain of a low-pass filter by putting $s = 0$ into the transfer function as well. If we put $s = 0$ into the $\frac{V_B(s)}{V_A(s)} = \frac{1}{0.0001s+1}$, the result will be 1. The dB equivalent of 1 is $20\log(1) = 0$ dB.

According to Fig. 6.26, the gain reaches to -3 dB at around $1.01 \times 10^4 \frac{\text{Rad}}{\text{s}} \approx 1.608$ kHz. Therefore, the cut-off frequency of this filter is 1.608 kHz. This filter passes the frequencies less than 1.608 kHz, i.e. [0, 1.608 kHz].

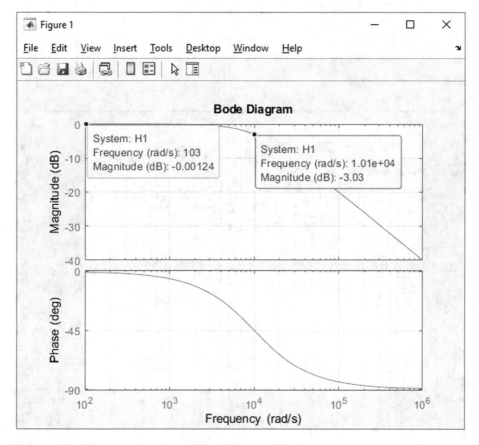

Fig. 6.26 Magnitude graph decreases to -3 dB around 10.1 kRad/s

In the code shown in Fig. 6.24 we had no control over the range of the drawn graph. The code shown in Fig. 6.27 draws the frequency response for [100 Hz, 10 kHz] range (Fig. 6.28).

```
Command Window                                    ⊙
>> s=tf('s');
>> H=1/(0.0001*s+1);
>> fmin=100;
>> fmax=10000;
>> wmin=2*pi*fmin;
>> wmax=2*pi*fmax;
>> w=logspace(log10(wmin),log10(wmax));
>> bode(H,w)
>> grid on
fx >>
```

Fig. 6.27 MATLAB code

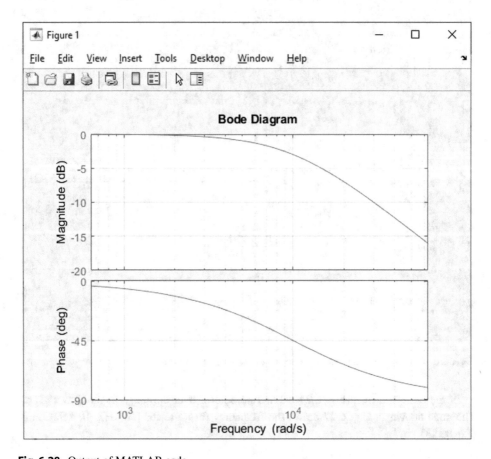

Fig. 6.28 Output of MATLAB code

You can change the unit of horizontal axis from Rad/s to Hz. with the aid of the technique shown in Sect. 6.4.

Let's study another example. The circuit shown in Fig. 6.29 is a low-pass filter as well. In Laplace domain circuit analysis, inductor L is replaced with impedance of Ls. The transfer function of circuit shown in Fig. 6.29 will be $V_B(s) = \frac{R_1}{R_1+L_1s} V_A(s) = \frac{1k}{0.01s+1k} V_A(s) \Rightarrow \frac{V_B(s)}{V_A(s)} = \frac{1000}{0.01s+1000}$. You can draw the frequency response of this filter and measure its cut-off frequency with MATLAB.

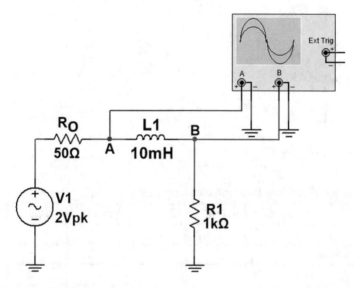

Fig. 6.29 Low-pass RL filter

6.5.2 Procedure

Make the circuit shown in Fig. 6.30 and fill the Table 6.3. Note that Ro shows the output resistance of function generator. Use MATLAB to draw the frequency response graph (i.e. graph of $\frac{Peak\ of\ Channel\ B}{Peak\ of\ Channel\ A}$ vs. frequency) and ensure that the circuit is a low-pass filter. You can use the Sect. A.7 of Appendix A if you don't know how to draw the frequency response graph with MATLAB.

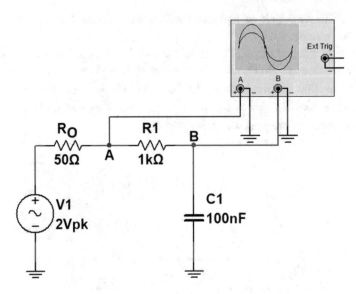

Fig. 6.30 Low-pass RC filter

Table 6.3 Frequency response for circuit shown in Fig. 6.30

Frequency	50 Hz	100 Hz	300 Hz	500 Hz	700 Hz	900 Hz
Peak of Channel B						
Peak of Channel A						
$\frac{PeakofChannelB}{PeakofChannelA}$						
Frequency	1100 Hz	1300 Hz	1500 Hz	1600 Hz	1700 Hz	
Peak of Channel B						
Peak of Channel A						
$\frac{PeakofChannelB}{PeakofChannelA}$						

Now change the circuit to what shown in Fig. 6.31. Fill the Table 6.4 and use MATLAB to draw the frequency response graph (i.e. graph of $\frac{Peak\ of\ Channel\ B}{Peak\ of\ Channel\ A}$ vs. frequency).

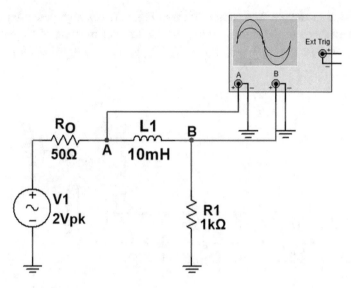

Fig. 6.31 Low-pass RL filter

Table 6.4 Frequency response for circuit shown in Fig. 6.31

Frequency	50 Hz	500 Hz	1 kHz	5 kHz	10 kHz	15.9 kHz	20 kHz	25 kHz	30 kHz
Peak of Channel B									
Peak of Channel A									
Peak of Channel B / Peak of Channel A									

The frequency which $\frac{Peak\ of\ Channel\ B}{Peak\ of\ Channel\ A} = \frac{1}{\sqrt{2}} = 0.707$ is called cut-off frequency of filter. In this experiment peak value of Channel A is around 2 V. So, we need to change the frequency until peak value of Channel B is around 1.41 V. Use this technique to measure the cut-off frequency of studied low-pass filters.

6.6 High-Pass Filter

6.6.1 Introduction

A high-pass filter is a circuit that passes signals with a frequency higher than a selected cutoff frequency and attenuates signals with frequencies lower than the cutoff frequency.

Let's study an example. The circuit shown in Fig. 6.32 is a high-pass filter (RO shows the output resistance of the signal generator). The transfer function of the circuit shown in Fig. 6.32 is $V_B(s) = \frac{R_1}{R_1 + \frac{1}{C_1 s}} V_A(s) = \frac{R_1 C_1 s}{R_1 C_1 s + 1} V_A(s) = \frac{0.0001 s}{0.0001 s + 1} V_A(s) \Rightarrow \frac{V_B(s)}{V_A(s)} = \frac{0.0001 s}{0.0001 s + 1}$.

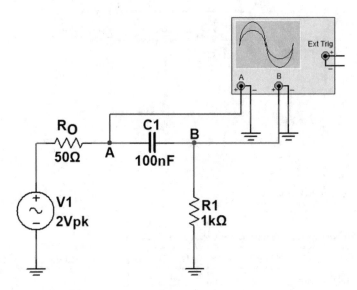

Fig. 6.32 High-pass RC filter

Frequency response of $\frac{V_B(s)}{V_A(s)} = \frac{0.0001 s}{0.0001 s + 1}$ can be drawn with the aid of the code shown in Fig. 6.33. Output of this code is shown in Fig. 6.34. Note that the Magnitude (dB) graph decreases as frequency decreases. This shows that the filter is high-pass.

Fig. 6.33 MATLAB code

```
Command Window
>> s=tf('s');
>> H=0.0001*s/(0.0001*s+1);
>> bode(H)
>> grid on
fx >>
```

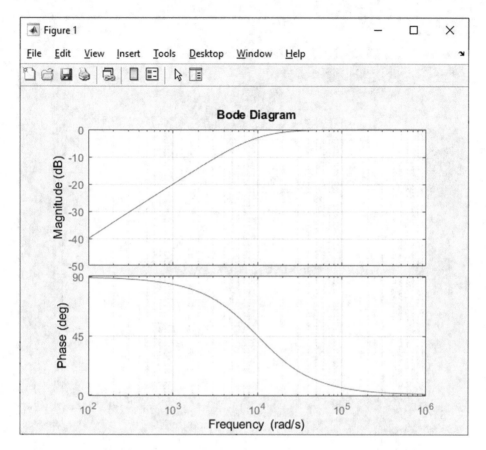

Fig. 6.34 Output of MATLAB code

The pass band gain of high-pass filters can be approximated by studying the behavior of the transfer function as s goes toward infinity. For $\frac{V_B(s)}{V_A(s)} = \frac{0.0001s}{0.0001s+1}$, $\lim_{s \to \infty}(\frac{0.0001s}{0.0001s+1}) = 1$ and dB equivalent of 1 is $20 \times log10(1) = 0$ dB.

Similar to the low-pass filter, the cut-off frequency is defined as the frequency which pass band gain decrease by 3 dB. The pass band gain is around 0 dB. According to Fig. 6.35, the gain reaches to -3 dB at around $1 \times 10^4 \frac{Rad}{s} \approx 1.592$ kHz. Therefore, the cut-off frequency of this filter is 1.592 kHz. This filter passes the frequencies bigger than 1.592 kHz, i.e. [1.592 kHz, ∞].

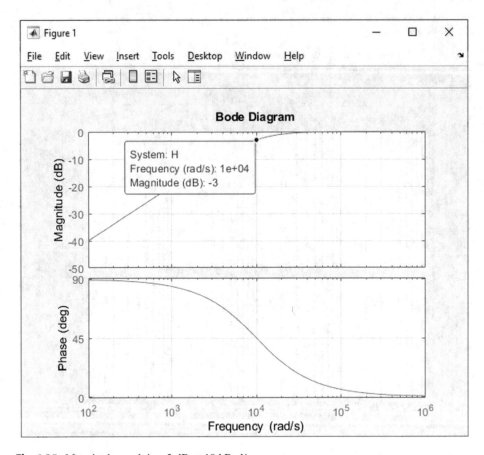

Fig. 6.35 Magnitude graph is −3 dB at 10 kRad/s

Let's study another example. The circuit shown in Fig. 6.36 is a high-pass filter as well. The transfer function of circuit shown in Fig. 6.36 will be $V_B(s) = \frac{L_1 s}{R_1 + L_1 s} V_A(s)$ $= \frac{0.01s}{0.01s + 1k} V_A(s) \Rightarrow \frac{V_B(s)}{V_A(s)} = \frac{0.01s}{0.01s + 1000}$.

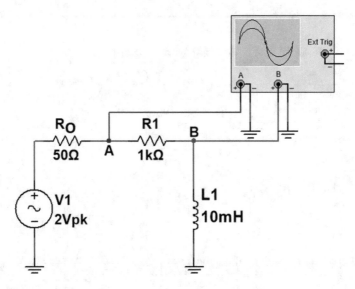

Fig. 6.36 High-pass RL filter

Let's draw the frequency response of circuits shown in Figs. 6.32 and 6.36 on the same graph. This permits us to compare the two graph easily. The code shown in Fig. 6.37 do this for us. Output of this code is shown in Fig. 6.38. According to Fig. 6.38, the circuit shown in Fig. 6.32 (RC circuit) has a lower cut-off frequency in comparison to the circuit shown in Fig. 6.36 (RL circuit).

```
Command Window                                    ⊙
>> s=tf('s');
>> H1=0.0001*s/(0.0001*s+1);
>> H2=0.01*s/(0.01*s+1000);
>> bode(H1,H2)
>> legend('RC circuit', 'RL circuit')
>> grid on
fx >>
```

Fig. 6.37 MATLAB code

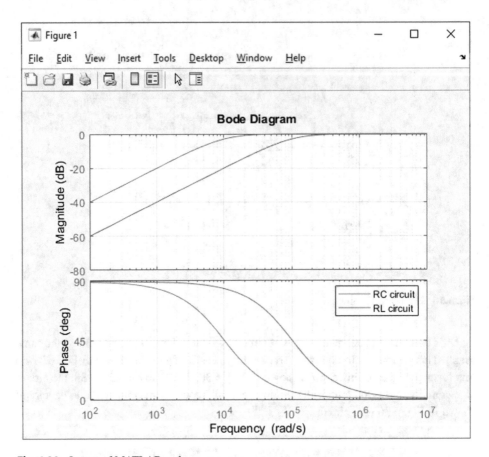

Fig. 6.38 Output of MATLAB code

6.6.2   Procedure

Make the circuit shown in Fig. 6.39 and fill the Table 6.5. Note that Ro shows the output resistance of function generator. Use MATLAB to draw the frequency response graph (i.e. graph of $\frac{Peak\ of\ Channel\ B}{Peak\ of\ Channel\ A}$ vs. frequency) and ensure that the circuit is a high pass filter. You can use the section A.7 of Appendix A if you don't know how to draw the frequency response graph with MATLAB.

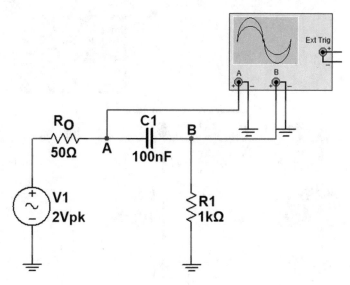

Fig. 6.39 High-pass RC filter

Table 6.5 Frequency response for circuit shown in Fig. 6.39

Frequency	150 Hz	1.5 kHz	1.6 kHz	2 kHz	4 kHz	6 kHz
Peak of Channel B						
Peak of Channel A						
$\frac{\text{Peak of Channel B}}{\text{Peak of Channel A}}$						

Frequency	8 kHz	10 kHz	15 kHz	20 kHz	25 kHz
Peak of Channel B					
Peak of Channel A					
$\frac{\text{Peak of Channel B}}{\text{Peak of Channel A}}$					

Now change the circuit to what shown in Fig. 6.40. Fill the Table 6.6 and use MATLAB to draw the frequency response graph (i.e. graph of $\frac{Peak\ of\ Channel\ B}{Peak\ of\ Channel\ A}$ vs. frequency).

Use the technique studied in previous experiment to measure the cut-off frequency of filters.

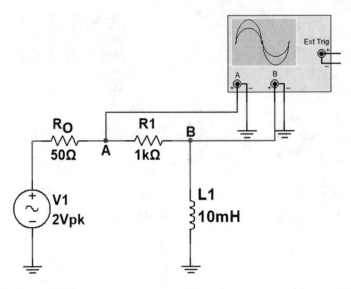

Fig. 6.40 High-pass RL filter

Table 6.6 Frequency response for circuit shown in Fig. 6.40

Frequency	500 Hz	5 kHz	10 kHz	15 kHz	15.9 kHz	20 kHz	25 kHz	30 kHz
Peak of Channel B								
Peak of Channel A								
Peak of Channel B / Peak of Channel A								

References for Further Study

1. Asadi F., Essential Circuit Analysis using Proteus, Springer, 2022.
2. Asadi F., Essential Circuit Analysis using LTspice, Springer, 2022.
3. Asadi F., Essential Circuit Analysis using NI Multisim and MATLAB, Springer, 2022.
4. Asadi F., Electric Circuit Analysis with EasyEDA, Springer, 2022.

Magnetic Coupling and Transformers

7

7.1 Introduction

This chapter studies the magnetic coupling and transformers. This chapter contains 3 experiments.

7.2 Dot Convention in Transformers

7.2.1 Introduction

For an ideal transformer with two windings (Fig. 7.1), the dot convention indicates the polarity of the windings.

Fig. 7.1 An ideal transformer

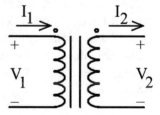

For voltage, when the excitation of one winding is at a positive maximum with respect to the dot, the voltage on the second winding will be at a positive maximum with respect to the dot. For current flow, the transformer obeys the law of conservation of power. Current flowing into the dot of one winding will cause a current to flow out of the dot on the other winding provided the transformer is transmitting a power flow.

© The Author(s), under exclusive license to Springer Nature Switzerland AG 2023
F. Asadi, *Electric Circuits Laboratory Manual*, Synthesis Lectures
on Electrical Engineering, https://doi.org/10.1007/978-3-031-24552-7_7

Assume that the transformer shown in Fig. 7.2 is given to you. In this experiment we want to find the location of dots.

Fig. 7.2 Location of dots are not known

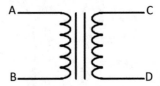

7.2.2   Procedure

Use a signal generator to generate a sinusoidal signal with maximum amplitude and frequency of 50 or 60 Hz. Connect the signal generator to primary windings and connect one of the secondary windings to ground (Fig. 7.3). Internal resistance of signal generator V1 is not shown in Fig. 7.3.

Fig. 7.3 Signal generator V1 is connected to the primary of the transformer

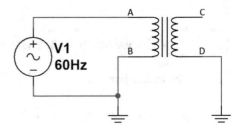

Connect the point A and C to an oscilloscope (Fig. 7.4) and pay attention to the phase difference between Channel A and B waveforms. Before connecting the node C to channel B of oscilloscope, use a DMM and measure the RMS value of voltage of node C. Multiply the read value with $\sqrt{2} = 1.41$ in order to find the peak value of voltage of node C. Ensure that this value is less than the maximum voltage that oscilloscope can handle safely.

Fig. 7.4 Voltage of nodes A
and C are monitored with an
oscilloscope

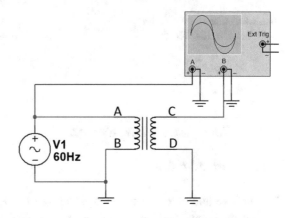

If 180° of phase difference exist between the two channels (Fig. 7.5), then the location
of dots is similar to Fig. 7.6.

Fig. 7.5 180° of phase
difference exists between the
two waveforms

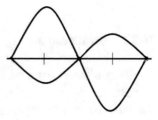

Fig. 7.6 Location of dots for
case shown in Fig. 7.5

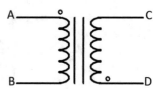

If there is no phase difference between the waveforms of the two channels (Fig. 7.7),
then the location of the dots is similar to Fig. 7.8.

Fig. 7.7 The two waveforms
are in phase

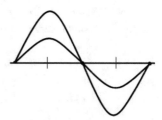

Fig. 7.8 Location of dots for
case shown in Fig. 7.7

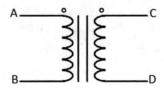

7.3   Turn Ratio of Transformer

7.3.1   Introduction

The voltage and current ratio of an ideal transformer is $\frac{V_s}{V_p} = \frac{I_p}{I_s} = \frac{N_s}{N_p}$, where $V_s =$ secondary voltage, $I_s =$ secondary current, $V_p =$ primary voltage, $I_p =$ primary current, $N_s =$ number of turns in the secondary winding and $N_p =$ number of turns in the primary winding.

In this experiment the technique to measure the turns ratio of a transformer is studied.

7.3.2   Procedure

Generate a sinusoidal signal with maximum amplitude and frequency of 50 or 60 Hz. Connect the signal generator to primary windings and leave the secondary open (Fig. 7.9). Internal resistance of signal generator V1 is not shown in Fig. 7.9. Use an AC voltmeter to measure the RMS of primary voltage. Call the measured value V1.

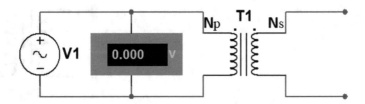

Fig. 7.9 Signal generator V1 is connected to primary winding

Now measure the RMS value of the secondary voltage (Fig. 7.10). Call the measured value V2. You can calculate the turn ratio with the aid of $\frac{V2}{V1} = \frac{N2}{N1}$.

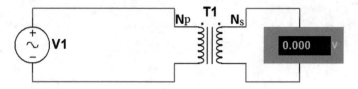

Fig. 7.10 AC voltmeter measures the secondary winding's voltage

7.4 Coupling Coefficient of Mutually Coupled Inductors

7.4.1 Introduction

In Fig. 7.11 two coupled inductors are connected together using the aiding method. Equivalent inductance for this connection is $L_{eq} = L_1 + L_2 + 2M$, where L_1, L_2 and M shows self-inductance of left coil, self-inductance of right coil and mutual inductance between the two coils, respectively.

Fig. 7.11 Equivalent inductance is $L_{eq} = L_1 + L_2 + 2M$

In Fig. 7.12 two coupled inductors are connected together using the opposing method. Equivalent inductance for this connection is $L_{eq} = L_1 + L_2 - 2M$. Definition of L_1, L_2 and M is similar to Fig. 7.11.

Fig. 7.12 Equivalent inductance is $L_{eq} = L_1 + L_2 - 2M$

The coefficient of coupling is defined as $k = \frac{M}{\sqrt{L_1 L_2}}$. If you connect the coils like Figs. 7.13 and 7.14, then the equivalent inductance will be $L_{eq} = \frac{L_1 L_2 - M^2}{L_1 + L_2 - 2M}$ and $L_{eq} = \frac{L_1 L_2 - M^2}{L_1 + L_2 + 2M}$, respectively.

Fig. 7.13 Equivalent inductance is
$L_{eq} = \frac{L_1 L_2 - M^2}{L_1 + L_2 - 2M}$

Fig. 7.14 Equivalent inductance is
$L_{eq} = \frac{L_1 L_2 - M^2}{L_1 + L_2 + 2M}$

7.4.2   Procedure

Prepare a coupled inductor (Fig. 7.15) or a transformer (Fig. 7.16). We want to measure the coupling coefficient between the two coils shown in Fig. 7.15 or primary and secondary windings shown in Fig. 7.16.

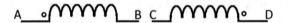

Fig. 7.15   Two magnetically coupled inductors

Fig. 7.16   Transformer

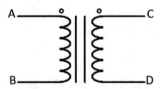

Connect the RLC meter to the left coil/primary winding (wire A and B in Figs. 7.15 and 7.16) and leave the right coil/secondary winding open. Write the value shown by the RLC meter and call it L_1.

Connect the RLC meter to the right coil/secondary winding (wire C and D in Figs. 7.15 and 7.16) and leave the left coil/primary open. Write the value shown by the RLC meter and call it L_2.

Connect one of the left coil/primary winding terminals to one of the right coil/secondary windings terminals. The connected two terminals are not important. For instance, all of the connections shown in Figs. 7.17, 7.18, 7.19 and 7.20 are acceptable. Connect the RLC meter to the remaining two terminals (A and C in Figs. 7.18 and 7.20 and A and D in Figs. 7.17 and 7.19). Write the value shown by the RLC meter and call it L_{eq}.

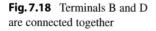
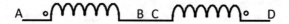

Fig. 7.17 Terminals B and C are connected together

Fig. 7.18 Terminals B and D are connected together

Fig. 7.19 Terminals B and C are connected together

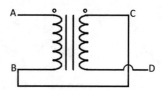

Fig. 7.20 Terminals B and D are connected together

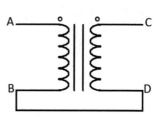

Compare the L_{eq} with $L_1 + L_2$. If $L_{eq} > L_1 + L_2$ then use the $L_{eq} = L_1 + L_2 + 2k\sqrt{L_1 L_2} \Rightarrow k = \frac{L_{eq}-L_1-L_2}{2\sqrt{L_1 L_2}}$ to calculate the coupling coefficient. If $L_{eq} < L_1 + L_2$ then use the $L_{eq} = L_1 + L_2 - 2k\sqrt{L_1 L_2} \Rightarrow k = \frac{L_1+L_2-L_{eq}}{2\sqrt{L_1 L_2}}$ to calculate the coupling coefficient.

When $L_{eq} > L_1 + L_2$ you make the connection similar to Figs. 7.18 or 7.19. When $L_{eq} < L_1 + L_2$ you make the connection similar to Figs. 7.17 or 7.20.

References for Further Study

1. Asadi F., Essential Circuit Analysis using Proteus, Springer, 2022.
2. Asadi F., Essential Circuit Analysis using LTspice, Springer, 2022.
3. Asadi F., Essential Circuit Analysis using NI Multisim and MATLAB, Springer, 2022.
4. Asadi F., Electric Circuit Analysis with EasyEDA, Springer, 2022.

Appendix A: Drawing Graphs with MATLAB®

A.1 Introduction

This appendix shows how to draw different types of graphs with MATLAB.

A.2 fplot Command

Assume that you want to draw the graph of $sin(x)$ for $[0, 2\pi]$ interval. The commands shown in Fig. A.1 do this job for you. Output of this code is shown in Fig. A.2.

Fig. A.1 fplot command can be used to draw symbolic expressions

```
Command Window
>> syms x
>> fplot(sin(x),[0,2*pi])
>> grid on
fx >>
```

© The Editor(s) (if applicable) and The Author(s), under exclusive license to Springer Nature Switzerland AG 2023
F. Asadi, *Electric Circuits Laboratory Manual*, Synthesis Lectures on Electrical Engineering, https://doi.org/10.1007/978-3-031-24552-7

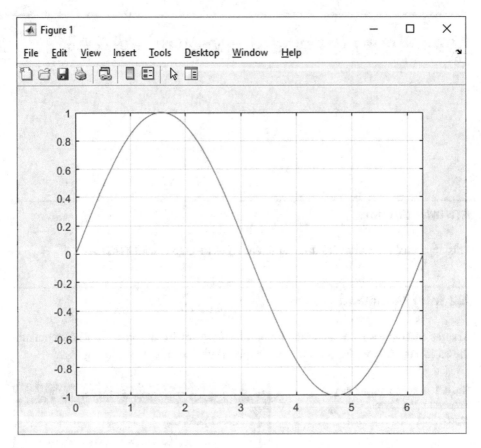

Fig. A.2 Output of code in Fig. A.1

You click on any point in order to read its coordinate (Fig. A.3).

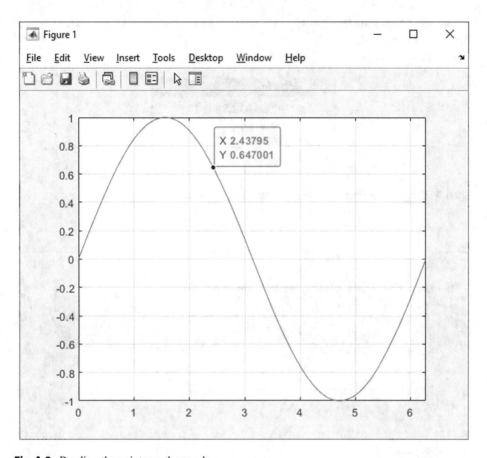

Fig. A.3 Reading the points on the graph

A.3 Plotting the Graph of a Numeric Data

In the previous section you learned how to draw the graph of a symbolic function. In this section we learn how to draw the graph of a numeric data. Plotting the graph of a numeric data is very simple in MATLAB. You need to use the plot command.

Let's make a numeric data (Fig. A.4).

Fig. A.4 Making a simple sample data

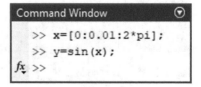

The plot command shown in Fig. A.5 draws the graph of the numeric data. Output of this code is shown in Fig. A.6.

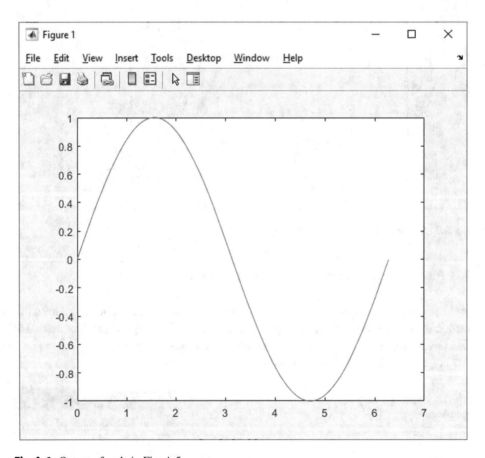

Fig. A.5 Drawing the graph of a numeric data with plot command

Fig. A.6 Output of code in Fig. A.5

You can add the operators shown in Tables A.1, A.2 and A.3 to produce more custom plots.

Table A.1 Types of lines

MATLAB command	Type of line
-	Solid
:	Dotted
-.	Dashdot
--	Dashed

Table A.2 Colors

MATLAB command	Color
r	Red
g	Green
b	Blue
c	Cyan
m	Magenta
y	Yellow
k	Black
w	White

Table A.3 Plot symbols

MATLAB command	Plot symbol
.	Point
+	Plus
*	Star
O	Circle
X	x-mark
s	Square
d	Diamond
v	Triangle (down)
^	Triangle (up)
<	Triangle (left)
>	Triangle (right)

Let's study an example. Values of voltage and current for a resistor is shown in Table A.4. We want to plot the graph of this data. We want to show the data points with circles and connect them together using dashed line with black color. The vertical axis and horizontal axis must have the labels "Current (A)" and "Voltage (V)", respectively. The title of the graph must be "I-V for a resistor". The commands shown in Fig. A.7 do what we need. The result is shown in Fig. A.8.

Table A.4 V-I values for resistor R1

V (volt)	I (Amper)
0.499	0.10
0.985	0.20
1.508	0.31
1.969	0.41
2.528	0.53
2.935	0.61
3.481	0.73
3.971	0.83
4.486	0.94
4.960	1.04
5.502	1.15
6.007	1.26
6.60	1.38

```
Command Window                                                                    ⊙
  >> V=[0.499 0.985 1.508 1.969 2.528 2.935 3.481 3.971 4.486 4.960 5.502 6.007 6.60];
  >> I=[0.1 0.2 0.31 0.41 0.53 0.61 0.73 0.83 0.94 1.04 1.15 1.26 1.38];
  >> plot(V,I,'--ko')
  >> title('I-V for a resistor')
  >> xlabel('Voltage(V)')
  >> ylabel('Current(A)')
  >> grid on
fx >>
```

Fig. A.7 Drawing the graph of data in Table A.4

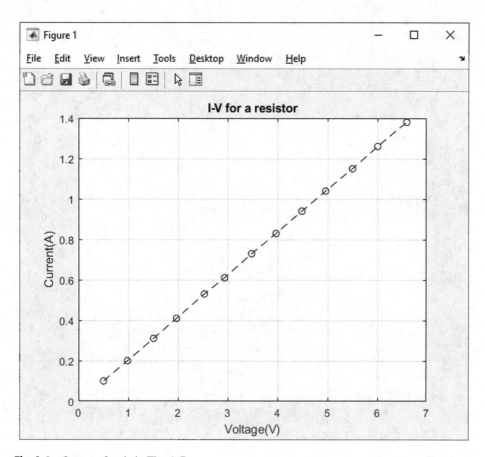

Fig. A.8. Output of code in Fig. A.7

The commands shown in Fig. A.9 draw the I-V graph of Table A.4. However, it uses red star for data points and solid black color for connecting the data points together (Fig. A.10).

```
Command Window
>> V=[0.499 0.985 1.508 1.969 2.528 2.935 3.481 3.971 4.486 4.960 5.502 6.007 6.60];
>> I=[0.1 0.2 0.31 0.41 0.53 0.61 0.73 0.83 0.94 1.04 1.15 1.26 1.38];
>> plot(V,I,'k',V,I,'r*'),grid minor
fx >>
```

Fig. A.9 Drawing the graph of data in Table A.4

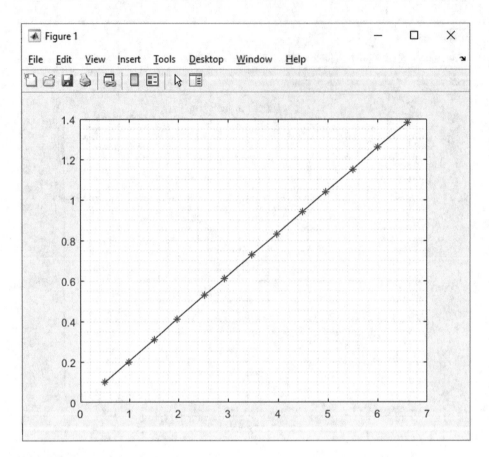

Fig. A.10 Output of code in Fig. A.9

A.4 Addition of Labels and Title to the Drawn Graph

Figure A.10 doesn't have any labels and title. You can add the desired labels and titles to it with the aid of insert menu (Fig. A.11).

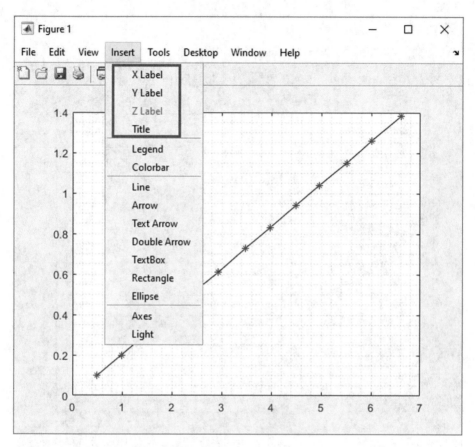

Fig. A.11   Addition of title and labels to the axis

A.5 Exporting the Drawn Graph as a Graphical File

You can copy the drawn graph to the clipboard easily with the aid of Edit>Copy Figure (Fig. A.12). After copying the graph to the clipboard you can easily paste it in programs like MS Word® by pressing Ctrl+V.

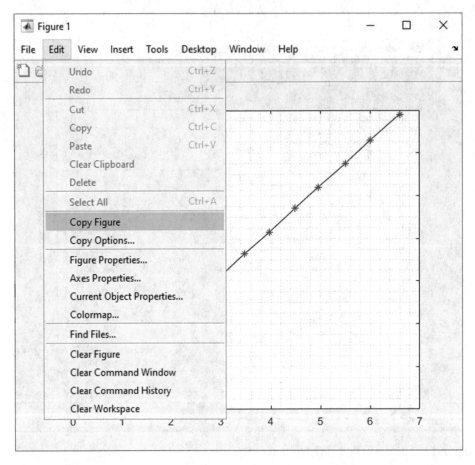

Fig. A.12 Copying the drawn figure to the clipboard memory

You can save the drawn graph as a graphical file as well. To do this, click use the File> Save As (Fig. A.13). After clicking, the save as window appears. Select the desired output format from the Save as type drop down list (Fig. A.14).

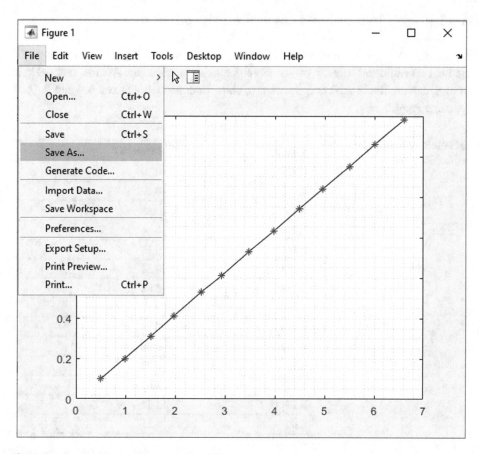

Fig. A.13 Saving the graph as a graphical file

File name:	untitled	∨
Save as type:	MATLAB Figure (*.fig)	∨

MATLAB Figure (*.fig)
Bitmap file (*.bmp)
∧ Hide Folders EPS file (*.eps)
Enhanced metafile (*.emf)
JPEG image (*.jpg)
Paintbrush 24-bit file (*.pcx)
Portable Bitmap file (*.pbm)
Portable Document Format (*.pdf)
Portable Graymap file (*.pgm)
Portable Network Graphics file (*.png)
Portable Pixmap file (*.ppm)
Scalable Vector Graphics file (*.svg)
TIFF image (*.tif)
TIFF no compression image (*.tif)

Fig. A.14 Selection of desired type of file

A.6 Drawing Two or More Graphs on the Same Axis

Sometimes you need to show two or more datasets on the same graph. You need to use the hold on command to show two or more graphs simultaneously. Assume that we have another dataset (Table A.5) and we want to show both datasets (Tables A.4 and A.5) on the same graph.

Table A.5 V-I values for resistor R2

V (volt)	I (Amper)
0.579	0.10
0.978	0.17
1.598	0.28
1.976	0.34
2.496	0.43
2.953	0.51
3.458	0.60
4.068	0.71
4.450	0.78
4.917	0.86
5.35	0.93
5.75	1.01
6.37	1.11
6.60	1.15

The commands shown in Fig. A.15 draws the graph of both datasets on the same graph. Output of this code is shown in Fig. A.16.

```
Command Window                                                                    ⊙
  >> V1=[0.499 0.985 1.508 1.969 2.528 2.935 3.481 3.971 4.486 4.960 5.502 6.007 6.60];
  >> I1=[0.1 0.2 0.31 0.41 0.53 0.61 0.73 0.83 0.94 1.04 1.15 1.26 1.38];
  >> V2=[0.579 0.978 1.598 1.976 2.496 2.953 3.458 4.068 4.450 4.917 5.35 5.75 6.37 6.60];
  >> I2=[0.1 0.17 0.28 0.34 0.43 0.51 0.60 0.71 0.78 0.86 0.93 1.01 1.11 1.15];
  >> plot(V1,I1,'b',V1,I1,'r*')
  >> hold on
  >> plot(V2,I2,'k',V2,I2,'r+')
  >> grid minor
  >> xlabel('Voltage (V)')
  >> ylabel('Current (A)')
  >> title('Comparison of I-V plot of R1 and R2')
fx >>
```

Fig. A.15 Drawing the graph of Tables A.4 and A.5 on the same graph

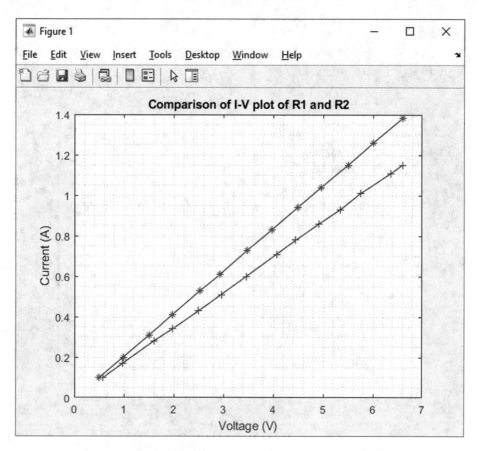

Fig. A.16 Output of code in Fig. A.15

You can use the Insert>Legend to show which graph belongs to which resistor (Fig. A.17).

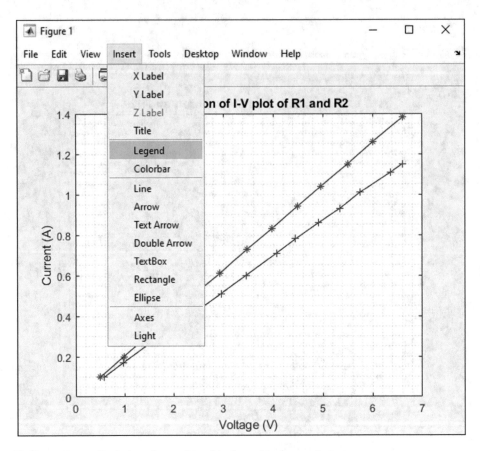

Fig. A.17 Insert>Legend can be used to add a legend to the graph

After clicking the Insert>Legend, the legend shown in Fig. A.18 will be added to the graph.

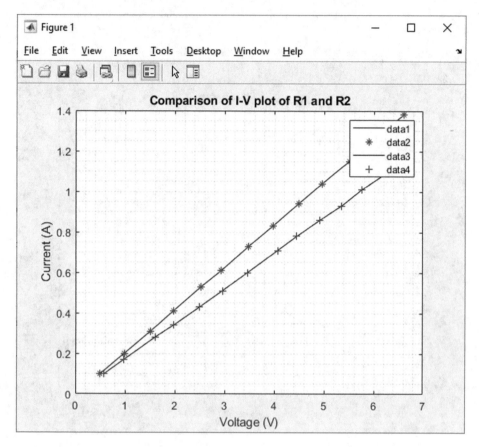

Fig. A.18 Legend is added to the graph

Double click the data1 in the legend box (Fig. A.18) and enter the desired text. Repeat this for data2, data3 and data4 in Fig. A.18. You can move the legend box by clicking on it, holding down the mouse button and dragging it to the desired location (Fig. A.19). You can even right click on the legend box and use the predefined locations (Fig. A.20).

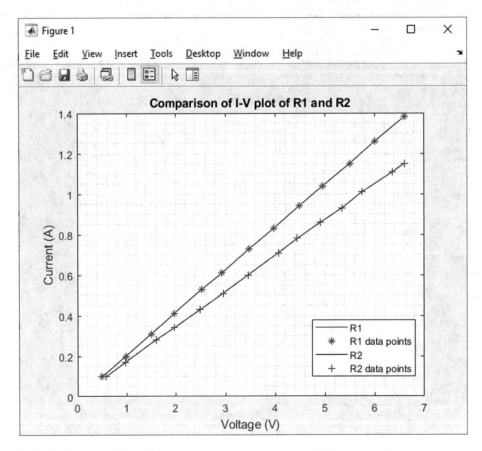

Fig. A.19 Customized legend

A.7 Logarithmic Axis

We used linear axis in order to draw the I-V graph of studied resistors. If you want to draw the frequency response graphs you need to use logarithmic axis. The linear axis is not a suitable option for frequency response graphs.

Fig. A.20 Default locations for legend

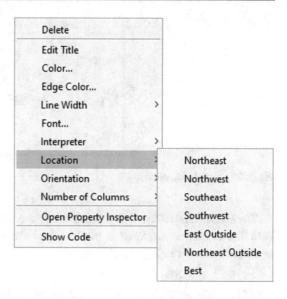

Let's study an example. Assume the frequency response given in Table A.6. This table shows the frequency response of the circuit shown in Fig. A.21.

Table A.6 Frequency response of a RC circuit

Frequency (Hz)	Magnitude $\left(\left\lvert \frac{V_o(j\omega)}{V_{in}(j\omega)}\right\rvert\right)$	Phase $\left(\sphericalangle \frac{V_o(j\omega)}{V_{in}(j\omega)}\right)$
1	1.000	$-0.36°$
10	0.998	$-3.60°$
20	0.992	$-7.16°$
50	0.954	$-17.44°$
100	0.847	$-32.13°$
150	0.728	$-43.30°$
200	0.623	$-51.48°$
250	0.537	$-57.51°$
300	0.469	$-62.05°$
350	0.414	$-65.54°$
400	0.370	$-68.30°$
450	0.333	$-70.51°$
500	0.303	$-72.34°$
550	0.278	$-73.85°$
600	0.256	$-75.14°$

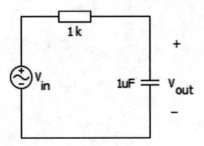

Fig. A.21 Simple RC circuit

The commands shown in Fig. A.22, draws the frequency response of the data in Table A.6. The command semilogx is used to draw the frequency response graph. Output of this code is shown in Fig. A.23.

```
Command Window
>> f=[1 10 20 50 100 150 200 250 300 350 400 450 500 550 600];
>> Amp=[1 0.998 0.992 0.954 0.847 0.728 0.623 0.537 0.469 0.414 0.370 0.333 0.303 0.278 0.256];
>> Phase=-[0.36 3.6 7.16 17.44 32.13 43.30 51.48 57.51 62.05 65.54 68.30 70.51 72.34 73.85 75.14];
>> subplot(211),semilogx(f,20*log10(Amp)),grid minor
>> title('Frequency responce of the RC circuit')
>> xlabel('Freq(Hz.)')
>> ylabel('Amplitude (dB)')
>> subplot(212),semilogx(f,Phase),grid minor
>> xlabel('Freq(Hz.)')
>> ylabel('Phase(Degrees)')
fx >> |
```

Fig. A.22 Drawing the graph of data in Table A.6

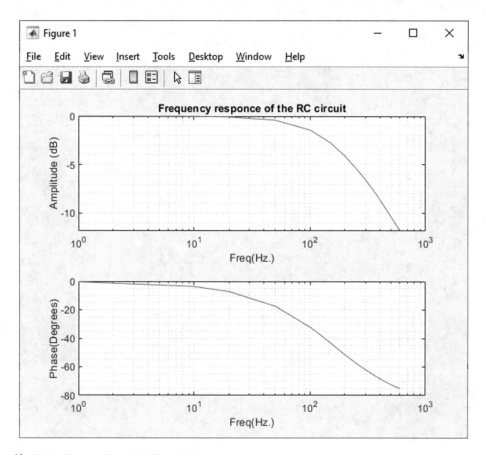

Fig. A.23 Output of code in Fig. A.22

Reference for Further Study

1. Asadi F., Applied Numerical Analysis with MATLAB®/Simulink®, Springer, 2022.

Appendix B: Root Mean Square

B.1 Introduction

An AC voltmeter/ammeter measures the RMS value of applied signal. RMS value has many applications in electrical engineering. This appendix reviews this import concept.

B.2 Root Mean Square (RMS) of a signal

Consider the simple circuit shown in Fig. B.1. The input source is a periodic voltage source, i.e., $v(t + T) = v(t)$. The load is purely resistive with value of R.

Fig. B.1 A resistor is connected to a periodic voltage source

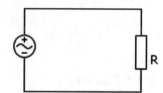

The average power consumed by the resistor is:

$$P = \frac{1}{T}\int_0^T p(t)dt = \frac{1}{T}\int_0^T v(t) \times i(t)dt = \frac{1}{T}\int_0^T \frac{v(t)^2}{R}dt = \frac{1}{R}\left[\frac{1}{T}\int_0^T v(t)^2 dt\right] \quad \text{(B.1)}$$

F. Asadi, *Electric Circuits Laboratory Manual*, Synthesis Lectures on Electrical Engineering, https://doi.org/10.1007/978-3-031-24552-7

Now consider the circuit shown in Fig. B.2. The input source is a constant DC voltage source, i.e., $v(t) = V_{dc}$.

Fig. B.2 The same resistor is
connected to a DC source

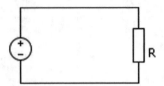

In this case the power consumed by the resistor is $\frac{V_{dc}^2}{R}$. Power consumption of both circuits are the same when $V_{dc} = \sqrt{\frac{1}{T}\int_0^T v(t)^2 dt}$. Since,

$$\frac{1}{R}\left[\frac{1}{T}\int_0^T v(t)^2 dt\right] = \frac{V_{dc}^2}{R} \Rightarrow V_{dc} = \sqrt{\frac{1}{T}\int_0^T v(t)^2 dt} \tag{B.2}$$

The $\sqrt{\frac{1}{T}\int_0^T v(t)^2 dt}$ is called Root Mean Square (RMS) or effective value of signal $v(t)$. So, RMS value of periodic signal $v(t)$ is a DC value which produce the same amount of heat in the resistive load as the periodic signal $v(t)$.

The RMS can be defined for the current waveforms as well.

$$I_{rms} = \sqrt{\frac{1}{T}\int_0^T i(t)^2 dt}. \tag{B.3}$$

B.2.1 Example 1

Determine the RMS value of the periodic pulse waveform shown in Fig. B.3.

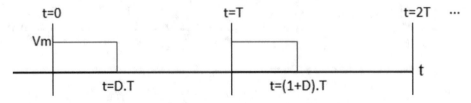

Fig. B.3 Waveform of Example 1

Solution

$$v(t) = \begin{cases} V_m & 0 < t < DT \\ 0 & DT < t < T \end{cases}$$

$$V_{rms} = \sqrt{\frac{1}{T}\int_0^T v(t)^2 dt} = \sqrt{\frac{1}{T}\left(\int_0^{DT} V_m^2 dt + \int_{DT}^T 0 dt\right)} = \sqrt{\frac{1}{T}(V_m^2 DT)} = V_m\sqrt{D}.$$

B.2.2 Example 2

Determine the RMS values of the following waveforms ($\omega = \frac{2\pi}{T}$).

(a) $v(t) = V_m \sin(\omega t)$.
(b) $v(t) = |V_m \sin(\omega t)|$.
(c) $v(t) = \begin{cases} V_m \sin(\omega t) & 0 < t < \frac{T}{2} \\ 0 & \frac{T}{2} < t < T \end{cases}$.

Solution

(a)

$$V_{rms} = \sqrt{\frac{1}{T}\int_0^T (V_m \sin(\omega t))^2 dt} = \sqrt{\frac{1}{T} \times V_m^2 \int_0^T \sin^2(\omega t) dt}$$

$$= \sqrt{\frac{V_m^2}{T}\int_0^T \frac{1 - \cos(2\omega t)}{2} dt} = \sqrt{\frac{V_m^2}{T}\int_0^T \frac{1}{2} dt - \int_0^T \frac{\cos(2\omega t)}{2} dt}$$

$$= \sqrt{\frac{V_m^2}{T} \times \left(\frac{T}{2} - \frac{\sin(2\omega t)}{4\omega}\right)\Big|_0^T} = \sqrt{\frac{V_m^2}{T} \times \frac{T}{2} - 0}$$

$$= \sqrt{\frac{V_m^2}{2}} = \frac{V_m}{\sqrt{2}}$$

(b) RMS value of $v(t) = |V_m \sin(\omega t)|$ is the same as $v(t) = V_m \sin(\omega t)$. Since

$(|V_m\sin(\omega t)|)^2 = (V_m\sin(\omega t))^2$. So, RMS value of $v(t) = |V_m\sin(\omega t)|$ is $\frac{V_m}{\sqrt{2}}$. Graph of $v(t) = |V_m\sin(\omega t)|$ is shown in Fig. B.4. Such a waveform is called Full Wave Rectified.

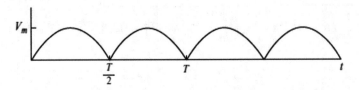

Fig. B.4 Full wave rectified sinusoidal waveform

(c) Graph of $v(t) = \begin{cases} V_m\sin(\omega t) & 0 < t < \frac{T}{2} \\ 0 & \frac{T}{2} < t < T \end{cases}$ is shown in Fig. B.5. Such a waveform is called Half Wave Rectified.

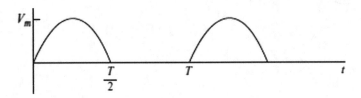

Fig. B.5 Half wave rectified sinusoidal waveform

$$V_{rms} = \sqrt{\frac{1}{T}\left(\int_0^{\frac{T}{2}}(V_m\sin(\omega t))^2 dt + \int_{\frac{T}{2}}^{T} 0\, dt\right)} = \sqrt{\frac{1}{T} \times V_m^2 \int_0^{\frac{T}{2}} \sin^2(\omega t)\, dt}$$

$$= \sqrt{\frac{V_m^2}{T}\int_0^{\frac{T}{2}}\frac{1 - \cos(2\omega t)}{2}dt} = \sqrt{\frac{V_m^2}{T}\int_0^{\frac{T}{2}}\frac{1}{2}dt - \int_0^{\frac{T}{2}}\frac{\cos(2\omega t)}{2}dt}$$

$$= \sqrt{\frac{V_m^2}{T} \times \left(\frac{t}{2} - \frac{\sin(2\omega t)}{4\omega}\right)\Big|_0^{\frac{T}{2}}} = \sqrt{\frac{V_m^2}{T} \times \frac{T}{4} - 0} = \sqrt{\frac{V_m^2}{4}} = \frac{V_m}{2}$$

RMS of triangular wave shapes can be calculated using the formulas shown in Fig. B.6.

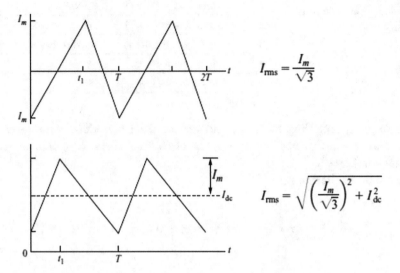

Fig. B.6 RMS value of triangular waveforms

B.3 Effective Value of Sum of Two Periodic Signals

Consider two periodic waveforms, i.e., $v_1(t + T) = v_1(t)$, $v_2(t + T) = v_2(t)$. The RMS value of sum of two waveforms ($v(t) = v_1(t) + v_2(t)$) is:

$$V_{rms}^2 = \frac{1}{T} \int_0^T (v_1 + v_2)^2 dt$$

$$= \frac{1}{T} \int_0^T (v_1^2 + 2v_1v_2 + v_2^2) dt = \frac{1}{T} \int_0^T v_1^2 dt + \frac{1}{T} \int_0^T 2v_1v_2 dt + \frac{1}{T} \int_0^T v_2^2 dt$$

$$(B.4)$$

Sometime the $\frac{1}{T} \int_0^T v_1(t)v_2(t)dt$ term is zero. The $\frac{1}{T} \int_0^T v_1(t)v_2(t)dt$ is the iner product of $v_1(t)$ and $v_2(t)$. When $\frac{1}{T} \int_0^T v_1(t)v_2(t)dt = 0$, the signals $v_1(t)$ and $v_2(t)$ are called orthogonal. Table B.1 shows some of the important orthogonal functions.

Table B.1 Some of the important orthogonal functions ($\omega = \frac{2\pi}{T}$, $n \neq m$ and k is a constant)

No.	$v_1(t)$	$v_2(t)$
1	$\sin(n \times \omega \times t + \varphi_1)$	$\sin(m \times \omega \times t + \varphi_2)$
2	$\sin(n \times \omega \times t + \varphi_1)$	$\cos(m \times \omega \times t + \varphi_2)$
3	$\cos(n \times \omega \times t + \varphi_1)$	$\cos(m \times \omega \times t + \varphi_2)$
4	$\sin(n \times \omega \times t + \varphi_1)$	k
5	$\cos(m \times \omega \times t + \varphi_1)$	k

For instance, according to the second row of the table, $\sin(n \times \omega \times t + \varphi_1)$ and $\cos(m \times \omega \times t + \varphi_2)$ (when $n \neq m$) are orthogonal since $\frac{1}{T} \int_0^T \sin(n\omega t + \varphi_1) \times \cos(m\omega t + \varphi_2)dt = 0$.

For orthogonal functions,

$$V_{rms}^2 = \frac{1}{T} \int_0^T (v_1 + v_2)^2 dt = \frac{1}{T} \int_0^T (v_1^2 + 2v_1 v_2 + v_2^2)dt$$

$$V_{rms}^2 = \frac{1}{T} \int_0^T v_1^2 dt + \frac{1}{T} \int_0^T 2v_1 v_2 dt + \frac{1}{T} \int_0^T v_2^2 dt$$

$$V_{rms}^2 = \frac{1}{T} \int_0^T v_1^2 dt + \frac{1}{T} \int_0^T v_2^2 dt$$

$$V_{rms} = \sqrt{V_{1,rms}^2 + V_{2,rms}^2} \qquad (B.5)$$

RMS value of sum of more than two orthogonal functions (each two terms are assumed to be orthogonal) can be calculated in the same way:

$$(v(t) = \sum_{n=1}^{N} v_n(t) \forall k, l 1 \leq k \leq N, o1 \leq l \leq N, ok \neq l, o\frac{1}{T} \int_0^T v_k(t)v_l(t)dt = 0) \Rightarrow$$

$$V_{rms} = \sqrt{V_{1,rms}^2 + V_{2,rms}^2 + V_{3,rms}^2 + \cdots} = \sqrt{\sum_{n=1}^{N} V_{n,rms}^2}. \qquad (B.6)$$

B.3.1 Example 3

Determine the RMS value of $v(t) = 4 + 8\sin(\omega_1 t + 10°) + 5\sin(\omega_2 t + 50°)$ under the following conditions.

(a) $\omega_2 = 2\omega_1$
(b) $\omega_2 = \omega_1$.

Solution

(a) When $\omega_2 = 2\omega_1$, the $v(t) = 4 + 8\sin(\omega_1 t + 10°) + 5\sin(2\omega_1 t + 50°)$. According to Table B.1, all the functions are orthogonal to each other, so

$$V_{rms} = \sqrt{V_{1,rms}^2 + V_{2,rms}^2 + V_{3,rms}^2} = \sqrt{4^2 + \left(\frac{8}{\sqrt{2}}\right)^2 + \left(\frac{5}{\sqrt{2}}\right)^2} = 7.78V$$

(b) When $\omega_2 = \omega_1$, the $v(t) = 4 + 8\sin(\omega_1 t + 10°) + 5\sin(\omega_1 t + 50°)$. $8\sin(\omega_1 t + 10°)$ and $5\sin(\omega_1 t + 50°)$ are not orthogonal to each other. So, we can't use the previous formullas. Note that $a \times \sin(\omega t) + b \times \cos(\omega t) = \sqrt{a^2 + b^2}\sin(\omega t + tan^{-1}(\frac{b}{a}))$. So,

$$v(t) = 4 + 8\sin(\omega_1 t + 10°) + 5\sin(\omega_1 t + 50°)$$
$$= 4 + 12.3\sin(\omega_1 t + 25.2°)$$

The two terms of last equation are orthogonal to each other (see Table B.1). So, the RMS is

$$V_{rms} = \sqrt{4^2 + \left(\frac{12.3}{\sqrt{2}}\right)^2} = 9.57V.$$

B.3.2 Example 4

In this example we show how RMS values can be calculated with the aid of MATLAB®. Assume $v(t) = 311\sin(2\pi \times 60t) + 100\sin(2\pi \times 2 \times 60t) + 20\sin(2\pi \times 3 \times 60t)$ is given. The RMS can be calculated easily:

$$V_{rms} = \sqrt{\left(\frac{311}{\sqrt{2}}\right)^2 + \left(\frac{100}{\sqrt{2}}\right)^2 + \left(\frac{20}{\sqrt{2}}\right)^2} = 231.43V$$

The commands shown in Fig. B.7 calculates the RMS value of given signal. The first two lines sample a period of given signal. The sampling time is $\frac{1}{6000} = 166.7\,\mu s$. The rms command is used to calculate the RMS value of sampled signal.

```
Command Window
    >> t=0:1/6000:1/60;
    >> v=311*sin(2*pi*60*t)+100*sin(2*pi*2*60*t)+20*sin(2*pi*3*60*t);
    >> rms(v)

ans =

    230.2829
fx >> |
```

Fig. B.7 Calculation of RMS value of $v(t)$ with $\frac{1}{6000}$ steps

The result is 230.283 which is a little bit lower than the expected value of 231.43. if you decrease the sampling time from $166.7\,\mu s$ to $16.67\,\mu s$ you get a more accurate result (Fig. B.8).

```
Command Window
    >> t=0:1/60000:1/60;
    >> v=311*sin(2*pi*60*t)+100*sin(2*pi*2*60*t)+20*sin(2*pi*3*60*t);
    >> rms(v)

ans =

    231.3158
fx >> |
```

Fig. B.8 Calculation of RMS value of $v(t)$ with $\frac{1}{60000}$ steps

B.4 Measurement of RMS of Signals

The cheap multimeters are not suitable devices to measure the RMS value of non-sinusoidal signals. The cheap multimeters are able to measure the RMS value of pure sinusoidal signals, i.e. the one shown in Fig. B.9.

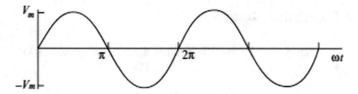

Fig. B.9 Pure sinusoidal waveform

Figure B.10 shows one of the methods that cheap multimeter uses measure the RMS of a signal. V_X is the signal under measurement. Assume that V_X is a pure sinusoidal waveform, i.e. a signal such as the one shown in Fig. B.9. Then the capacitor is charged up to Vm Volts (voltage drop of diode is neglected) where Vm is the peak value of voltage under measurement. So, Analog-to-Digital converter reads the maximum of input signal. The read value is simply multiplied by $\frac{1}{\sqrt{2}}$, and the result, i.e., $\frac{Vm}{\sqrt{2}}$, is the RMS value of input signal. This method only works for pure sinusoidal signals and doesn't produce correct result if the input signal is not pure sinusoidal.

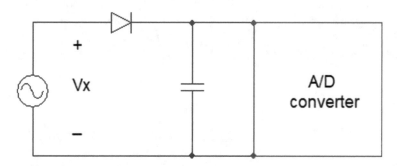

Fig. B.10 A simple circuit for detection of input AC signal peak value

The expensive multimeters samples the input waveform and uses a processor to calculate the RMS value. So, the wave shape of input signal doesn't affect the measurements. Such a multimeter has "TRUE RMS" label on it. So, ensure that your multimeter is TRUE RMS type if you want to measure the RMS of a non-sinusoidal signal. Digital oscilloscopes can be used to measure the RMS of non-sinusoidal signals as well.

Reference for Further Study

1. Asadi F., Simulation of Power Electronics Circuits with MATLAB®/Simulink®, Springer, 2022.

Printed in the United States
by Baker & Taylor Publisher Services